LIVING AND FOSSIL
BRACHIOPODS

Biological Sciences

———

Editor
PROFESSOR A. J. CAIN
MA, D. PHIL
Professor of Zoology
at the University of Liverpool

LIVING AND FOSSIL

BRACHIOPODS

M. J. S. Rudwick

Sedgwick Museum, Cambridge
Fellow of Corpus Christi College

HUTCHINSON UNIVERSITY LIBRARY
LONDON

HUTCHINSON & CO (*Publishers*) LTD
178-202 Great Portland Street, London W1

London Melbourne Sydney
Auckland Johannesburg Cape Town
and agencies throughout the world

First published 1970

*This book has been set in Times type, printed in Great Britain
on smooth wove paper by Anchor Press, and
bound by Wm. Brendon, both of Tiptree, Essex*

ISBN 0 09 103080 3 (cased)
0 09 103081 1 (paperback)

CONTENTS

Preface   9

1 INTRODUCTION AND GENERAL FEATURES

   Brachiopods and evolutionary theory   13
   General features of brachiopods   16
   Classification of brachiopods   21
   Evolutionary history of brachiopods   24
   Appendix: Outline classification of
      Brachiopoda   26

2 SHELL AND MANTLE

   Function of the shell   29
   Mode of growth of the shell   30
   Composition of the shell   34
   Structure of the mantle   37
   Structure of the shell   39
   Mantle caecae   42

3 MUSCLES AND HINGES

   The hinge   50
   Articulation   51
   Interlocking structures   55
   Muscular leverage   56
   Muscle platforms   62
   Leverage in inarticulates   68

4 RELATION TO SUBSTRATE

Structure of the pedicle   74
Attachment of the pedicle   77
Pedicle foramen   78
Cementation   85
Free-lying brachiopods   87
Swimming brachiopods   91
Quasi-infaunal brachiopods   91
Burrowing brachiopods   94

5 SENSES AND PROTECTION

The snapping reaction   98
The setae   100
Sensory spines   107
Zig-zag slits   110

6 LOPHOPHORE AND FEEDING

The ciliary pump   117
Feeding and rejection mechanisms   120
Digestion and defaecation   122
Excretory system   123
Apertures   124
Supporting structures   125
Growth of the lophophore   130
Simple lophophores   132
Spiral lophophores   134
Terebratulide lophophores   140
Lobate lophophores   143
Rhythmic feeding mechanisms   145

7 REPRODUCTION AND ECOLOGY

The gonads   152
Brood pouches   153
Larval development   155
Rate of growth and population structure   157
Ecological distribution   157
Biotic relations   161

Contents

8 HISTORY OF THE PHYLUM

Pre-Cambrian origins 164
Early Cambrian radiation 164
Ordovician radiation 165
Siluro–Devonian expansion 168
Late Palaeozoic expansion 170
Permo–Triassic extinction 172
Mesozoic and Cenozoic faunas 173

9 ASPECTS OF EVOLUTION

Affinities to other phyla 175
Origin of the Cambrian fauna 176
Adaptive radiation 177
The problem of extinction 181
Brachiopods and bivalves 182
Conclusion 184

References 185

Index 193

PREFACE

Brachiopods are abundant and important fossils, but living species are rare. Most research on brachiopods has been the work of palaeontologists, many of whom have been more interested in using brachiopods for geochronology than in reconstructing their modes of life or tracing their evolution. Even when palaeontologists have attempted biological interpretations their conclusions have often been vitiated by a lack of knowledge and understanding of living brachiopods. In an attempt to narrow this 'information gap', I have tried in this book to integrate as closely as possible our present knowledge of living brachiopods with that of fossil brachiopods. In doing so I have tried to bear in mind the very different background knowledge of neontologists and palaeontologists, and to avoid unnecessary technicalities. This book is not a detailed or comprehensive specialist review, but a personal attempt to synthesise current knowledge of brachiopods from a particular viewpoint. This viewpoint is summarised by the term 'functional morphology'.

A general synthesis of the biology of living and fossil brachiopods is especially worthwhile at the present time. A great mass of detailed work on brachiopods has been published in the past ten years; some of it is of high quality, but little of it has found its way into textbooks or teaching courses. Inevitably my treatment of the subject is coloured by my own interests, and I have discussed some aspects more fully than others. This is not due wholly to personal bias, however; some topics have had to be treated briefly simply because the necessary detailed research on them has still to be done. For example, very little worthwhile work has yet been done on the ecology or palaeo-ecology of brachiopods, which should be almost the centre of their 'natural history'. Likewise, there is still no comparative account of the functional morphology of muscle and hinge systems, although these features have been regarded as especially important for classification. Moreover, all too many papers on systematic palaeontology still give no adequate descriptions or illustrations of morphology, which makes it difficult to interpret their findings in biological terms unless one has access to the original specimens.

This work first took shape as an advanced course of lectures in palaeozoology, and I am grateful to Professor O. M. B. Bulman, F.R.S., for having given me this initial opportunity to clarify my ideas. A draft in book form was later written as an essay for the Sedgwick Prize at Cambridge. But I then delayed any revision for publication until I could incorporate the information contained in the section on Brachiopoda (Part H) in the *Treatise on Invertebrate Palaeontology*,[58] and especially its comprehensive review of the hitherto chaotic systematics of the phylum. This long-awaited work at last became available in this country early in 1966. It would be difficult to overestimate its value and importance as a compilation at the present time; yet it is a regrettable symptom of the divorce between neontology and palaeontology that even the *Treatise* contains two successive chapters, entitled 'Anatomy' and 'Morphology', which contain *separate* discussions of living and fossil brachiopods respectively. Nevertheless, its summary of the characters of some 2000 genera has been invaluable for the interpretation of the functional evolution of the phylum, and especially for the location of crucial points that required more detailed study; and I have drawn heavily on its taxonomic scheme, though with some important modifications. On the interpretative level, both functionally and phylogenetically, my conclusions often differ from those of the *Treatise*. Although the completion of the present book has been unavoidably delayed during the period since the publication of the *Treatise*, and although many new taxa have been erected in that time, I have decided in general to retain the *Treatise* definitions of genera and families, since these were at least formulated under a relatively uniform editorial policy.

It would not be possible in a book of this length to try to substantiate every assertion with published authorities. To keep the bibliography within bounds, I have confined references (with the exception of those used as sources for illustrations, and a few others) to those published in the last decade: most of the relevant papers of earlier date can be located from references in the *Treatise*, or in Hyman's valuable chapter on living brachiopods in *The Invertebrates*.[51] In addition to published sources, Dr Richard Cowen has very kindly allowed me to use some of the unpublished results of his work, and I have also incorporated much of my own unpublished research.

Many of the diagrams in this book are new; the rest are re-drawn from previously published figures or photographs. Some have merely been re-drawn to conform in style to the rest; others have been adapted in ways which involve a greater or lesser degree of functional interpretation, for which of course I must be held responsible. In such cases the author of the original drawing or photograph is cited in acknowledgement of its source, but may not necessarily agree with the interpretation I have given to his material.

One of the practical disadvantages of writing about the functional morphology of fossils is that the reasons for arriving at a particular interpretation cannot be set down briefly. Since functions cannot be directly observed in operation in fossil animals, the morphology has to be considered in detail from the viewpoint of all competing interpretations. Such detailed analyses would be out of place in a book of this kind; if included, they would expand it into an entirely different work. But the penalty for their exclusion is that any interpretations that are summarised briefly are bound to appear dogmatic, or at least unsupported and no better than many alternative interpretations. All I can say here is that most of the interpretations given in this book have been tested according to a method which I have analysed elsewhere,[79] and are based on detailed study, though some more tentative suggestions are made in the hope that others may test them more rigorously.

The research on which this book is based would have died an early death had it not been for several zoologist friends, and especially the late Professor Carl Pantin, F.R.S., who encouraged my enthusiasm for studying the functional morphology of fossils at a time when even that phrase seemed to be unknown to the majority of invertebrate palaeontologists, and when any such straying from the strait and narrow path of taxonomic palaeontology was liable to be dismissed as pointless or—worse still—as 'merely speculative'. To Dr Richard Cowen I am indebted for the stimulus of an enthusiasm for brachiopods as immoderate as my own, and for much unstinted hard work in assistance with detailed research. Most of the illustrations in this book are the work of Mrs Jenny Friend, who has been extremely patient in interpreting my draft sketches and converting them into attractive finished drawings. Without her assistance this book would never have been completed, or would have had to be published with far fewer illustrations.

I am indebted above all to the Science Research Council of Great Britain, for a grant which has enabled me to enlarge the scale of my research in the last four years, and which has made it possible to attempt a broad synthesis of adaptive evolution in the brachiopods.

I hope that this book will persuade both neontologists and palaeontologists that the Brachiopoda offer peculiarly attractive advantages as subjects for research into some of the fundamental problems of the history and development of life. It will have served its purpose if it stimulates them to test more thoroughly and critically the interpretations I have made.

I

INTRODUCTION AND GENERAL FEATURES

Brachiopods and evolutionary theory

Brachiopods are not generally familiar animals, and neither their Latin name nor the English term 'lamp-shells' means very much to most people. To explain that they are 'a kind of shell-fish but really quite different from other shell-fish' is still not very helpful, though it does at least convey the fact that they are aquatic animals with a hard external shell. Even among zoologists they are not well known, and most textbooks of systematic zoology give them merely a page or two in a chapter on 'minor phyla'. Yet on turning to any textbook of palaeontology we find that this minor phylum has become 'major', and dozens of pages may be devoted to a detailed description of what now appear to be highly complicated objects equipped with a most formidable terminology. But at the end of this they are likely to remain no more than 'objects', and the reader may still feel he has little idea what these fossils were like when they were living animals.

There is perhaps no other phylum in which this contrast of approach is so marked. There are other groups that are of great significance to the student of living animals but of relatively little interest to the palaeontologist, simply because their remains are rarely preserved in the fossil state. But with the brachiopods this position is reversed. They are of great importance to the palaeontologist, for their remains are often abundant in fossil-bearing rocks of almost every geological age; yet few kinds have survived to the present day, and few zoologists have ever seen them alive. The general lack of communication between neontologists and palaeontologists is particularly unfortunate here, because it tends to obscure from both groups the high biological interest of this phylum. Brachiopods may seem of minor importance if we limit our view to the present evolutionary 'instant' in which we can study animals alive. But in relation to the total history of the biosphere the Brachiopoda must certainly be classed as one of the major phyla of invertebrate animals.

Among such major groups the phylum has several features that make it especially suitable for a study of the broader aspects of the evolutionary process. It has an exceptionally long recorded history: indeed no group with a satisfactory record of preservation has a longer history. The first known brachiopods are found in some of the earliest Lower Cambrian strata, and some are alive today. This represents a time-span of nearly six hundred million years. Moreover, between the earliest and the latest the continuity of the record is also exceptionally good, indeed almost unrivalled. Brachiopods can be found in variety and often in profusion, at least somewhere in the world, in rocks of almost every geological age from Cambrian to Cenozoic. Although the most widely known brachiopod genus (*Lingula*) is a textbook example of evolutionary stability, the phylum as a whole was far from static during this lengthy period. Not only did some groups rise slowly to dominance while others waned, but at times there were 'bursts' of rapid diversification, and at other times massive extinctions. Indeed it is this continual change in the detailed and in the overall character of the phylum that makes the study of brachiopods so important for the dating and correlation of strata. Of course, like every other fossil group, brachiopods are affected by the fragmentary nature of the fossil record; but it is fragmentary more in detail than in its broader features. It is very difficult to find satisfactory evidence of small-scale (inter-specific) evolutionary changes; but the broader course of events is often relatively clear, and gives exceptionally good opportunities for studying the processes of supra-generic evolution. Moreover, the growth-stages of many structural features are 'built in' to each brachiopod specimen, so that there is some possibility of relating the history of the individual (ontogeny) to that of the race (phylogeny).

However, a full fossil record is not by itself enough. Any meaning-

---

Fig. 1. A community of living brachiopods on a rock surface. There is a terebratulide with a large smooth shell (*Waltonia*), a rhynchonellide with a ridged (costate) shell (*Notosaria*), and a terebratulide with a very small shell (*Pumilus*); all three species are represented by juvenile and adult individuals. There is an abundant associated fauna of other epifaunal suspension-feeding animals, including sponges, hydroids, tube-worms, bryozoans and bivalves.[76] Drawn from photographs of a boulder from a rock-pool near Christchurch, New Zealand.[86]

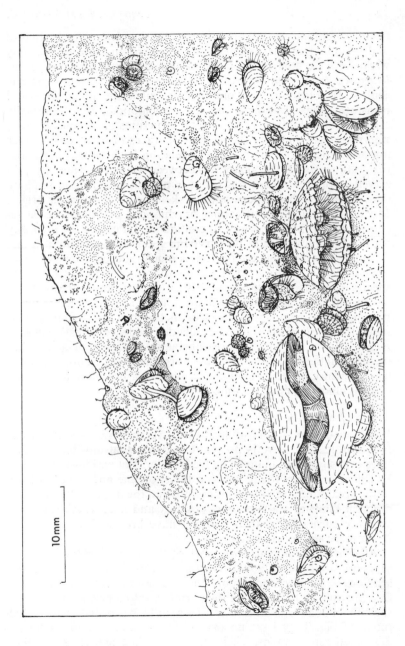

10mm

ful interpretation of evolutionary patterns involves some biological understanding of the organisms as they are 'in themselves': only when we have seen each of them—at least in the mind's eye—as *living* organisms, and not merely as objects with a certain structure, shall we be able to see the evolutionary significance of their structural transformations. But here too the Brachiopoda offer unusually attractive possibilities. Their skeletal structures are frequently very well preserved, and bear the imprint or traces of many of the other important organs of the body. The relation between the 'hard' and the 'soft' parts of the anatomy—between what is commonly preserved in the fossil state and what is not—is much more intimate than in most other groups with a skeleton of ectodermal origin. Much of the anatomy of extinct brachiopods can therefore be reconstructed with a fair degree of confidence. Of course the main key for this is a knowledge of the surviving species: here again, it is just fortunate that although the phylum is much reduced from its former abundance it is still not wholly extinct, and the range of surviving species is sufficient to provide a fairly satisfactory key to the anatomy of the extinct species. But the living species also provide at least a partial key to the adaptations of extinct species; they enable us to make the most of the sometimes slender evidence of past environments, and of the much more substantial evidence of adaptation recorded in the structure of the fossil organism itself. This gives us at least a chance of interpreting the evolution of brachiopods in terms of their changing modes of life.

General features of brachiopods

Brachiopods are 'a kind of shell-fish'. All the living species are in fact marine, and there is no fossil evidence that any brachiopods have ever succeeded in invading fresh water. But various species can be found living anywhere between the shoreline and the deep floor of the oceans, between the polar regions and the tropics, and on many different kinds of sea-floor, rocky, sandy and muddy. All the fossil evidence suggests that their habitats have been at least as varied, if not more so, in the past.

All living brachiopods form part of the sessile benthonic fauna, and almost all are epifaunal. They are permanently attached to the sea-floor or to objects on it; only during a short larval period are they free to swim, or to drift with the currents, into a new area. Most of their functional organisation can be seen to be related to this fixed mode of life. They have no organs of locomotion and no highly developed sense organs. Their defence against predators or other dangers is entirely passive: all they can do, apart from being incon-spicuous, is to close their shell when danger threatens. They cannot go actively in search of food, but must utilise whatever source of food approaches them: hence, like many other sessile animals, they feed by filtering small particles out of the sea-water that surrounds them. Reproduction, too, generally involves no more than shedding the genital products into the water, and relying on the chances of fertilisation and larval development there.

Some extinct brachiopods seem to have escaped from one or other of these limiting features of the mode of life of living brachiopods, yet few of them deviated very far. But within this severely restricted way of life, brachiopods have been able to develop a surprising diversity. Nearly two thousand genera and probably about thirty thousand species have already been described and named; and since many of these seem to have been very local in distribution, there were probably many more which have not been preserved by the chances of geological history. Even allowing for the fact that many species have been founded on very minute differences (some of which would probably be regarded as intra-specific if they were found in a living population), such a large number of genera and species does indicate the great diversity that has developed within a basically rather stable structural organisation. The equivalent (homologous) parts of the brachiopod body can generally be recognised without much difficulty in different brachiopods. We do not see within the phylum the kinds of radical transformation of anatomy and physio-logy that distinguish, for example, the mammals from the fish among the vertebrates or the cephalopods from the bivalves among the molluscs. But what we can see, displayed by an exceptionally fine fossil record, is the remarkable diversity of form—and apparently of habit too—that can develop *within* the severe limitations imposed by a basically unchanging anatomical and physiological organisation. In this lies much of the evolutionary interest of the phylum.

Like other 'shell-fish', the *shell* of a brachiopod is its most con-spicuous part. Like bivalve molluscs such as cockles and mussels, the shell in fact consists of two separate parts, the *valves*, which between them enclose almost all the rest of the organism. Thus an important function of the shell—perhaps its primary function—is that it pro-vides the organism with protective armour. The shell almost always shows perfect bilateral symmetry; but the plane of symmetry runs through the two valves and not, as in most bivalve molluscs, between them. This serves as a key to the conventional orientation of a brachiopod, which is an essential aid to description (Fig. 2). The plane of bilateral symmetry is regarded as the vertical longitudinal plane, separating the left half of the animal from the right. The second axis is provided by the differences between the two valves: one is termed *dorsal* (D.V. on text-figures) and the other *ventral* (V.V.), though these directions have no necessary relation to the attitude adopted by the living animal. When the shell is open, the valves gape apart more widely at one end than at the other, where they may be hinged to one another. This provides the third axis for orientation: the widely gaping end is termed *anterior*; the other end, where there may be a hinge, is the *posterior*.

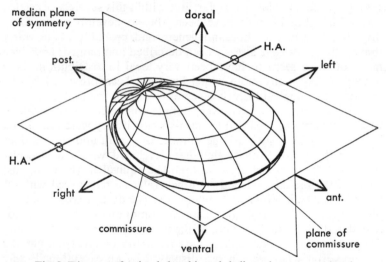

Fig. 2. Diagram of a simple brachiopod shell, to show conventional orientation. Dorsal valve above. H. A., hinge axis. Some arbitrarily chosen growth-lines and radial lines are also shown.[71]

The presence or absence of a hinge is the most important character by which the two main divisions of the phylum can be recognised. In the *INARTICULATA* the valves are not rigidly hinged to one another; in the *ARTICULATA* there is a hinge of interlocking *teeth* and *sockets* between the valves. The hinge restricts the valves to simple opening and closing movements, whereas the valves of an inarticulate have greater freedom of movement. There are therefore corresponding differences in the system of *muscles* by which these movements are effected. Some extinct articulates lost their hinge teeth and sockets in the course of evolution, but there is never any risk of mistaking them for inarticulates.

The shell of a brachiopod gives a misleading idea of the real size of the organism. Much of the space enclosed by the valves, the *shell cavity*, is strictly speaking 'outside' the brachiopod (Fig. 3). The real 'body' of the animal occupies only a small part of the shell-cavity on its posterior side. From the body, two thin sheets of living tissue, the *mantle lobes*, extend forwards and line the inner surface of the two valves. But between them is a large *mantle cavity* filled with sea-water and communicating directly with the exterior whenever the valves gape apart.

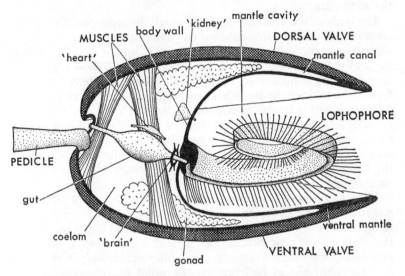

Fig. 3. Generalised anatomy of a brachiopod: highly diagrammatic sagittal section. [86]

The mantle cavity contains a large and complicated organ termed the *lophophore*. Like the shell, and indeed like almost every organ of the brachiopod, the lophophore shows perfect bilateral symmetry. It consists of a pair of feathery *brachia*, which project from the body at the back of the mantle cavity and are variously twisted up within the cavity. (It was at one time thought that these feathery 'arms' could be uncoiled, protruded out of the shell, and possibly even used for locomotion like the foot of a bivalve mollusc; this was the origin of the term Brachiopoda, i.e. 'arm-footed'.) The lophophore is primarily a feeding organ, serving to filter small particles of food out of the sea-water; but it also plays an important role in respiration

and other metabolic activities. In some articulate brachiopods the
lophophore is partly supported on an internal skeleton, the *brachidium*, which is attached to the dorsal valve. (For this reason the dorsal
valve is sometimes termed the 'brachial valve'; but the name is not a
good one, since it is not directly applicable to species without
brachidia, and in any case the brachia themselves do not 'belong' to
one valve more than the other.) The form of the brachidium gives
some indirect evidence of the form of the lophophore in extinct
brachiopods, and is also important in the high-level classification of
the Articulata.

The body of the brachiopod, from which the lophophore projects,
contains not only the muscles that move the valves but also a simple
digestive system, *excretory* and *reproductive* organs, and the central
parts of simple *circulatory* and *nervous* systems. All these organs are
suspended in a fluid-filled cavity, the *coelom*, which is separated
from the mantle cavity by a thin *body wall*. The coelom, and some of
the organs within it, also extend into the mantle lobes, and may leave
more or less direct traces on the shells of fossil species. There are no
highly developed sense organs, and the sensitivity of the animal
seems to be virtually confined to the edges of the mantle lobes, where
the living tissues have their most direct contact with the external
environment.

Most brachiopods are attached to the sea-floor by a stalk, the
pedicle, which projects between the valves on the posterior side of the
shell. In fact two quite different organs are included under this name:
the pedicles of inarticulates and articulates develop from quite
different sources during ontogeny, and differ significantly in structure
and function. But in both groups the point of emergence of the
pedicle is often incorporated in the ventral valve as a *pedicle foramen*.
(For this reason the ventral valve is sometimes termed the 'pedicle
valve'; but the term is open to the same objections as 'brachial
valve', since it cannot be applied to all brachiopods except by
analogy, and the pedicle does not—except in the inarticulates—
'belong' to one valve more than the other.) The pedicle foramen is
preserved in fossil brachiopods and is important in the high-level
classification of the phylum. A few brachiopods have lost the pedicle
altogether—if they ever had one—and are attached by the ventral
valve itself, which is cemented to the substrate: this habit was much
more common among extinct brachiopods. Some fossil species seem
to have lost the pedicle without acquiring any alternative mode of
attachment, and must have been free-living.

Classification of brachiopods

The classification of brachiopods has remained extremely confused until recently, owing to attempts to use one or a few isolated morphological characters as diagnostic features of major taxa. A more nearly natural, phylogenetic classification is now emerging from the attempt to group the taxa together 'from the bottom upwards'. Although clear-cut formal diagnoses cannot easily be given for any of the major taxa, most brachiopod genera can be assigned without difficulty to one or another of the Orders into which the two classes are divided (Fig. 4).

Among inarticulates there are two major Orders. The LINGULIDA (lingulides) mostly have shells of elliptical outline, with gently and equally convex valves. There is no pedicle foramen, and the pedicle, when present, must have emerged posteriorly between the valve edges. In the ACROTRETIDA (acrotretides), on the other hand, the shells are more commonly circular in outline, with valves of unequal and sometimes strong convexity. The pedicle generally emerged through a foramen in the ventral valve. This Order also includes an important group (craniaceans) in which the pedicle is replaced by a cemented ventral valve. Apart from these two Orders, there are also some small but highly interesting groups in Cambrian strata, which are so different from any other inarticulates that separate Orders have been erected for them (obolellides, paterinides, kutorginides).

The earliest articulates, which are probably ancestral to all the rest, are the ORTHIDA (orthides). They generally have simple shells with gently convex valves, fine radial ridges on the external surface, and a broad hinge line on the posterior margin. The PENTA-MERIDA (pentamerides) usually have much more strongly convex shells, often smooth externally, and with conspicuous internal platforms for the attachment of the muscles; they are probably derived from orthide ancestors. The RHYNCHONELLIDA (rhynchonellides) also have strongly convex shells with no apparent hinge-line on the posterior margin; but they generally lack the internal platforms of pentamerides and almost always have strong radial ridges externally; like some pentamerides they have a very simple form of brachidium, and they may be derived from that group. The ATRYPIDA (atrypides) have shells similar to the rhynchonellides, from which they probably evolved, but have a complex spirally coiled brachidium. The same kind of brachidium is also characteristic of the SPIRIFERIDA (spiriferides); but in other respects, especially the broad hinge-line on the posterior margin, this group is closer

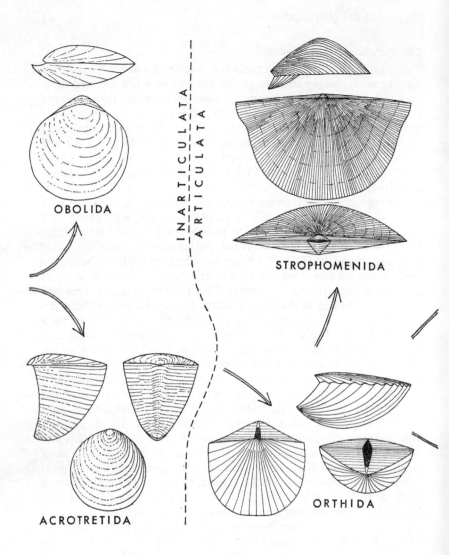

Fig. 4. Drawings of typical members of the main Orders of Brachiopoda, to illustrate general character of shells and probable evolutionary relations. Each shell is shown in standard dorsal, right lateral, and posterior or anterior views.[86]

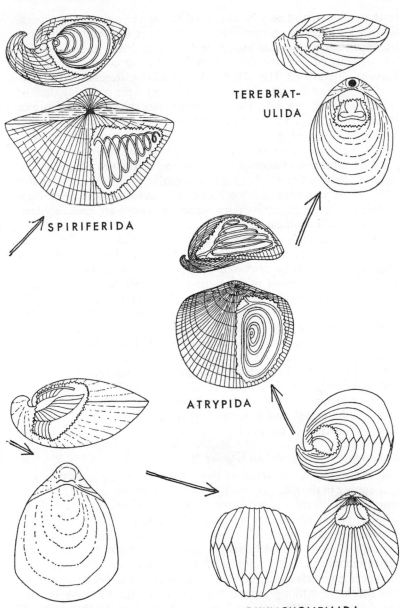

SPIRIFERIDA

TEREBRAT-
ULIDA

ATRYPIDA

PENTAMERIDA

RHYNCHONELLIDA

to the orthides, and may be unrelated to the atrypides. (The separation of atrypides from spiriferides is the most important difference between the classification used in this book and that of the *Treatise*; but at least in part the distinction follows the common usage of Russian literature.) The TEREBRATULIDA (terebratulides) resemble some of the atrypides externally, and were probably derived from them, but internally the brachidium has the form of a more or less complex loop. The terebratulides were the last Order to appear in the fossil record, and are also the most abundant at the present day. Finally, most of the STROPHOMENIDA (strophomenides) are quite different in appearance from any of the preceding Orders, because they generally have an externally concave dorsal valve which gives the whole shell a saucer-like or even cup-like form; their shells also have a distinctive microstructure. They were probably derived from orthide ancestors.

These descriptions are necessarily vague and generalised; but some of the distinctions between the Orders, and their constituent Superfamilies, will be discussed in greater detail in the rest of the book. For the purpose of exposition, it is necessary to refer to a general phylogeny of the brachiopods, although the justification for this particular scheme cannot be discussed adequately within the limits of this work (Fig. 99). Although some aspects of it are controversial, most of its broader features would, I believe, have fairly general agreement among those who have studied the phylum as a whole. In any case, in whatever way the constituent taxa are shuffled and re-arranged, it is almost certainly impossible to reach any scheme which avoids the multiple origin of many of the most important morphological and functional features. This will be a major theme of the rest of the book.

Evolutionary history of brachiopods

The brachiopods, like most of the other fossilisable animal phyla, first appear in the fossil record in Lower Cambrian strata. They were then already clearly distinct from any other phylum, and are without any but purely speculative ancestors. They were also already fairly diverse. Throughout the Cambrian period, inarticulates were commoner and more varied than the articulates, but neither class was very abundant until the subsequent Ordovician period. Both classes then became much more abundant and diverse, but this development was far more striking among the articulates than among the inarticulates; the latter never afterwards formed more than a small minority in brachiopod faunas, and were soon reduced to a few stable forms which have survived with little change to the present day. The articulates, on the other hand, underwent their most important phase of high-level evolution during the Ordovician period, and it is possible to recognise here the roots of most of the later diversification of the phylum. The many Superfamilies into which the six articulate Orders are divided had varying fortunes during the remainder of the Palaeozoic era. Various groups increased in importance, or were virtually replaced by others as dominant elements of the brachiopod fauna, some becoming completely extinct. But the articulates as a whole remained extremely abundant, and formed a very important element in the bottom-living marine faunas of the world. Even towards the end of the Palaeozoic, during the Permian period, brachiopods were still very abundant and diverse, and indeed included some of the most remarkable forms known in the history of the phylum.

But at about the end of the Palaeozoic era the Brachiopoda, in common with many other phyla of marine invertebrates, suffered the greatest crisis in their history. Although the Orders were not equally affected, the phylum as a whole never fully recovered from this crisis, and brachiopods have never again contributed more than a minor element to the overall benthonic fauna. Since early in the Mesozoic era the phylum has been fairly stable, however, and today rather less than three hundred species are known to survive.

Evolution is commonly described—with the vertebrates chiefly in mind—as though it were an irresistibly 'upward' or 'progressive' process. The Brachiopoda, as a phylum whose greatest abundance and diversity lie far back in the past, force us to reconsider this 'image' of the evolutionary process. Why, after successive phases of diversification and periods of great abundance throughout the Palaeozoic era, did they suffer such a severe crisis? Why did they fail to recover? In what sense, except by a circular definition, can their few living survivors be termed more 'successful' or 'better adapted' than the far more abundant and varied faunas of earlier periods? What 'progress', if any, is discernible within the pattern of their evolution? These are some of the questions that a study of the biology of living and fossil brachiopods may help to answer.

Three taxonomic levels are given, class, order, superfamily; suborders and families are omitted. A number in brackets indicates the number of genera described in each group in the *Treatise*;[58] the number recorded in Lower Cambrian, and the number extant (Recent), are also shown. The same classification is used for Fig. 99 and the evolutionary charts based upon it (Figs. 15, 33, 49, 63, 88).

Range

INARTICULATA

    LINGULIDA

Lingulacea (46)	L Camb (3) – Rec (2)
Trimerellacea (5)	M Ord – U Sil

    ACROTRETIDA

Acrotretacea (32)	L Camb (7) – Dev
Discinacea (11)	M Ord – Rec (3)
Siphonotretacea (5)	U Camb – Ord
Craniacea (15)	L Ord – Rec (3)

    OBOLELLIDA (5)   L Camb (5) – M Camb

    PATERINIDA (7)   L Camb (5) – M Ord

    KUTORGINIDA[a] **(3)**   L Camb (2) – ?M Camb

ARTICULATA

    Eichwaldiacea (4) (Order uncertain)   M Ord – Perm

    ORTHIDA

Billingsellacea (8)	L Camb (4) – L Ord
Orthacea (82)	L Camb (1) – U Dev
Enteletacea (66)	L Ord – U Perm
Clitambonitacea[b] (23)	L Ord – U Ord

STROPHOMENIDA

Plectambonitacea (53)	L Ord – U Dev
Strophomenacea (68)	L Ord – U Dev
Davidsoniacea (29)	L Ord – U Trias
Thecideacea[c] (13)	U Perm – Rec (2)
Triplesiacea[d] (10)	L Ord – U Sil
Chonetacea (34)	L Sil – L Jur
Koninckinacea[e] (6)	M Trias – L Jur
Strophalosiacea[f] (48)	L Dev – U Perm
Productacea (131)	L Dev – U Perm
Lyttoniacea (15)	U Carb – U Perm

PENTAMERIDA

Porambonitacea (40)	M Camb – L Dev
Pentameracea (44)	U Ord – U Dev
Stenocismatacea[g] (11)	M Dev – U Perm

RHYNCHONELLIDA

Rhynchonellacea[h] (259)	M Ord – Rec (11)

ATRYPIDA[i]

Atrypacea (42)	M Ord – U Dev
Dayiacea (13)	M Ord – M Dev
Retziacea[j] (16)	U Sil – U Trias
Athyridacea (41)	U Ord – U Trias

SPIRIFERIDA[k]

Cyrtiacea[l] (40)	L Sil – L Jur
Spiriferacea (89)	L Sil – U Perm
Spiriferinacea (19)	L Carb – L Jur
Reticulariacea (36)	L Dev – U Perm

TEREBRATULIDA

Stringocephalacea (39)	L Dev – Perm
Dielasmatacea (42)	L Dev – U Trias
Terebratulacea (90)	U Trias – Rec (12)
Zeilleriacea[m] (26)	L Dev – L Cret
Terebratellacea (51)	U Trias – Rec (36)

Notes: this classification is based on the *Treatise*, but with modifications as follows: [a]Kutorginida here transferred from 'class uncertain' to Inarticulata, in view of lack of true articulation; [b]Clitambonitacea here includes Gonambonitacea (10), which differ chiefly in shell structure; [c]Thecideacea transferred from 'order uncertain' to Strophomenida, in view of generally davidsoniacean morphology,[84] notwithstanding anomalous shell structure;[105] also includes Bactryniidae (1) transferred from Lyttoniacea on account of thecideacean morphology; [d]Triplesiacea here transferred from Orthida to Strophomenida in view of strophomenide shell structure;[105] [e]Koninckinacea here transferred from Spiriferida to Strophomenida in view of generally chonetacean morphology, and includes closely similar Cadomellacea (1),[35] notwithstanding their spiralia and anomalous shell structure;[105] [f]Strophalosiacea here includes similar though more aberrant Richthofeniacea (6); [g]Stenoscismatacea here transferred from Rhynchonellida to Pentamerida in view of pentameride-like muscle platforms, notwithstanding some rhynchonellide features;[45] [h]Rhynchonellacea here includes Rhynchoporacea (1), differing only in shell structure; [i]Atrypida as here defined includes part of 'Spiriferida', viz. Atrypidina, Retziidina, Athyrididina; [j]Retziacea here includes Athyrisinacea (5), differing only in shell structure; [k]Spiriferida as here defined includes only Spiriferidina; [l]Cyrtiacea here includes Suessiacea (12), differing chiefly in shell structure; [m]Zeilleriacea here includes similar but Palaeozoic Cryptonellacea (4).

2

SHELL AND MANTLE

Function of the shell

The shell is the most conspicuous part of a living brachiopod; in a fossil it is all that is normally preserved, even though it may bear traces of many other organs of the body. The form and function of the shell is therefore an appropriate point at which to begin the study of the biology of brachiopods.

The primary function of the shell seems to be that of providing the animal with protective armour. For a bivalved shell to fulfil this function effectively, it is necessary that the valves should be wholly external to the living tissues of the body, and that their edges should be able to fit tightly together. In this way they can seal all the living tissues from direct contact with the external environment. This 'specification' is fulfilled accurately by the shells of all living brachiopods. The shell is a purely external skeleton: the mineral and organic substances of which it is composed are secreted by the epithelial cells of the mantle that lines it internally. When the shell is closed the edges of the valves fit very tightly together along the *commissure* of the shell; only the pedicle remains partly exposed outside the shell (Fig. 5A).

The shells of most fossil brachiopods are essentially similar. If their 'soft parts' are reconstructed by homology with living species, it is clear that they too conformed closely to the same ideal 'specification' for a protective bivalved shell. It is therefore reasonable to say that protection has always been the main function of the shell in brachiopods.

A few late Palaeozoic strophomenides (richthofeniid strophalosiaceans and lyttoniaceans) appear to have lost this primary function of the shell, for the dorsal valve is very thin and delicate, and is recessed within a much larger and more robust ventral valve. The ventral mantle lobe must have extended beyond the edge of the dorsal valve in order to secrete the peripheral parts of the ventral valve, and was almost certainly exposed permanently to the external environment (Fig. 5B, C).[73, 104] The shell must have been much less effective

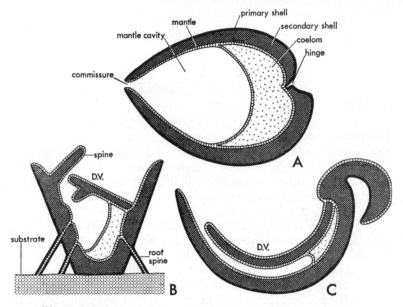

Fig. 5. Relations of shell and mantle in brachiopods, shown by diagrammatic longitudinal sections. A, a 'normal' brachiopod (e.g. any living species); B, a richthofeniid (aberrant strophalosiacean);[73] C, a lyttoniacean.[104] Primary shell black; secondary shell stippled; mantle tissue (observed in A, inferred in B and C) shown by cross-hatching.

than usual as protective armour, but the dorsal valve may have retained or acquired other functions (p. 145). These groups of highly aberrant brachiopods do not seem to be related to one another, and therefore this loss of the protective function of the shell may have occurred more than once, in the late Carboniferous period (Fig. 15).

Mode of growth of the shell
Since the mantle tissue is present only on the inner side of the shell, each valve can only grow in size by the accretion of new material contributed by the epithelial cells at the extreme edges of the mantle

lobes, which are closely adherent to the extreme edges of the valves (Fig. 9). All that the remainder of the epithelial cells can do is to thicken and strengthen the valves by adding new material internally. The entire *external* form of the shell is determined by the way in which successive increments are added to the edges of the valves during ontogeny.[71, 85] Minor irregularities between these increments often show on the external surfaces of the valves as *growth-lines*. By studying the growth-lines, it is quite easy to reconstruct all the stages by which the external form of the shell grew during ontogeny (Fig. 6).

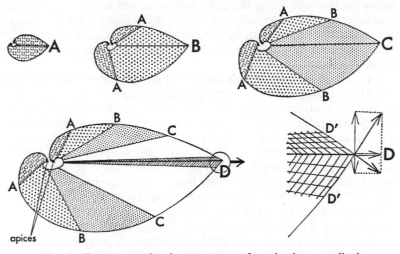

Fig. 6. Four stages in the ontogeny of a simple generalised brachiopod (right lateral view), illustrating the preservation of early growth-stages in the full-grown shell; protegula black. The enlargement of the anterior valve edges at stage D' to D illustrates the components of the relative growth-rates responsible for the maintenance of a logarithmic spiral form throughout the ontogeny of the shell.[85]

Each valve begins in the larval stage as a tiny plate termed a *protegulum*. (In *Lingula* there is initially only a single plate, which later splits into dorsal and ventral protegula;[113] but this is probably not universal among living inarticulates,[69] and it seems unwise to assume that it is a fundamental character distinguishing the inarticulates from the articulates.) The form of each valve thereafter depends simply on the relative rates at which new material is added in different directions around its edges. Each valve is in effect a disc of shell material, and each growth increment is a ring of material added to the periphery of the disc. A plane disc is in fact the simplest of a range of possible shapes, and will be formed if the valve grows straight outwards from the protegulum at an equal rate all round. If, as is more commonly the case, the growth includes a component directed towards the opposite valve, the valve will develop as a limpet-like cone, with the protegulum preserved at the *apex* of the valve. The height of the cone will depend simply on the ratio between the rate of growth towards the opposite valve and the rate of growth outwards. If the former of 'vertical' component is directed *away* from the opposite valve, the cone will be turned inside out, so that the valve becomes externally concave. It is possible for this component to change in 'sign' during the growth of the valve: for example, the valve can be convex near the apex but concave peripherally (Fig. 24).

In most actual brachiopods the rate of growth is much more rapid anteriorly than posteriorly; consequently the apex is generally much nearer the posterior margin of the shell, and may project posteriorly as a prominent 'beak' or *umbo*. Commonly the rate and direction of growth anteriorly, relative to that posteriorly, remains fairly constant during ontogeny; then the valves in lateral profile approximate to the form of logarithmic (equiangular) spirals, like many other accretionary structures. But in many other brachiopods the rates of growth changed significantly during ontogeny; and the young stages may be strikingly different in form from the adults, and difficult to recognise unless the growth-lines are compared. Failure to use the evidence of growth lines may even result in an assignment of juvenile and adult individuals to different species or even different genera!

Since the valve edges normally remain in close contact throughout ontogeny, their 'outward' rates of growth must necessarily be related, but their 'vertical' components can be entirely independent (Fig. 6). Hence the form and convexity (or concavity) of one valve may be very different from those of the other valves (Fig. 8). The bilateral symmetry of the shell, which is generally almost perfect, reflects an equality in the growth rates on the left and right halves of the valve edges. A few genera (e.g. *Uncites, Meekella*) are almost always symmetrical, the beaks of the valves having a twisted appearance: this is due to a slight inequality in the growth on left and right sides.

In the simpler kinds of shell the commissure lies in a single *commissural plane*, which is at right angles to the *median plane* of symmetry (Fig. 2, 7A). But if the growth on a particular *arc* of the valve edge is accelerated or retarded, or changed in direction, relative to that on the rest of the valve edge, it will produce a local *deflection* of that valve edge, which will be matched by a complementary deflection on the other valve edge. Such deflections generally develop over a considerable period of ontogeny (Fig. 7B), and are therefore preserved as features of the surface of the shell as well as its commissure. For example a simple *median deflection*, which is a very common feature in many groups of brachiopods, generally leaves as its 'growth-track' a median ridge or 'fold' on one valve surface and a corresponding groove or 'sulcus' on the other (Fig. 7C). Similarly, if the commissure becomes crinkled into a *serial deflection* during ontogeny, the valve surfaces will be corrugated into a series of ridges (*costae*) and grooves radiating from the apices (Fig. 62). Like the

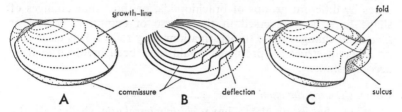

Fig. 7. Diagrams to show how a shell with a plane commissure (A) may be modified by a median fold and sulcus (C), if the commissure is affected progressively during ontogeny by the development of a median deflection (B). B represents commissures at successive growth-stages of C.[85]

overall shell shape, deflections generally show almost perfect bilateral symmetry, but a few genera (e.g. *Streptis, Rhactorhynchia*) have persistently asymmetrical deflections.

In principle it would be possible to describe completely the external shape of any brachiopod shell, however complex, in terms of the pattern of relative growth rates that operated on different parts of the two valve edges during the successive phases of ontogeny. Such a description, which for most actual brachiopods would be extremely complex, would be a purely formal analysis of the shape. The functional or adaptive significance of the shape would be an entirely separate question. But the possibility of such a formal analysis has highly important implications for evolutionary studies of brachiopods. An appropriate pattern of growth rates would be capable of producing *any* shape whatever that is inherently possible for a bivalved shell. It is in fact possible to program a computer to construct a matrix of finely graded variant forms, merely by giving

B

appropriate instructions for the variation of the basic growth rates and their ratios.[66]

The variety of shapes possible within such a multi-dimensional matrix is very great; and although brachiopods are almost entirely confined to that part of the matrix which represents bilaterally symmetrical forms, the actual shapes known in living and fossil brachiopods are remarkably diverse (Fig. 8). But it is perfectly possible—in principle—that the shape characteristic of any one species could have changed into the shape characteristic of any other species, simply by an appropriate modification of the growth rates. Such an evolutionary change could have happened once or many times, and would be perfectly reversible. It is therefore not surprising to find good evidence that many distinctive shapes have in fact been evolved several times by different groups of brachiopods, or that some changes of shape have been reversed in the course of evolution. Whatever *functional* changes these transformations may reflect, it is clear that no inherent *structural* problems would have been involved in their accomplishment.

Composition of the shell

Most of the inarticulates have *chitinophosphatic* shells; all the articulates have *calcareous* shells. In a chitinophosphatic shell a variable proportion of the material (42% in *Lingula*, 11% in *Discinisca*) is organic, being composed of true chitin and protein in roughly equal amounts. The inorganic material is mostly calcium phosphate, with small amounts of other inorganic salts. In a calcareous shell there is very little organic material, and none of it is chitin; nearly all the inorganic material is calcium carbonate, almost always in the form of calcite.[53] Calcite is more resistant to diagenetic change than aragonite (which characterises many molluscs), and this is partly responsible for the fine state of preservation of many fossil brachiopods. But perhaps the most revealing mode of preservation is one which involves the complete post-mortem replacement of the calcite by silica: for although this destroys the internal fine structure, it enables the shells to be extracted by acid from even hard limestones, revealing to perfection many delicate features which would otherwise be poorly known. Chitinophosphatic shells can often be similarly extracted, even if they are not silicified, by the use of weak acids which do not corrode the phosphate. The amino-acid assemblages in the shell proteins of living brachiopods seem to run parallel to the classification based on gross morphology, but in fossil specimens this phylogenetic information is obscured by diagenetic changes.[54]

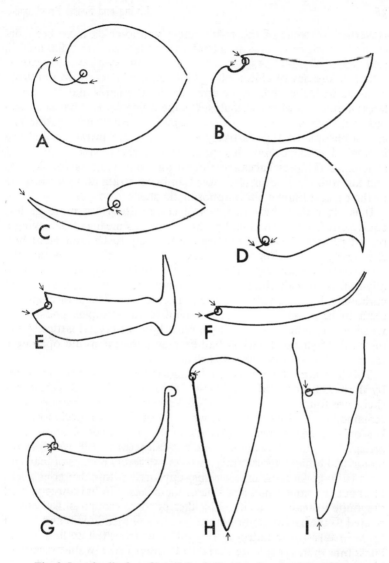

Fig. 8. Longitudinal profiles of the shells of various brachiopods, to illustrate diversity of form; not to scale. (A, pentameride *Conchidium*; B, spiriferide *Ambocoelia*; C, terebratulide *Terebrirostra*; D, rhynchonellide *Glossinotoechia*; E–I all strophomenides, *Leptaena, Rafinesquina, Auloprotonia, Geyerella* and *Richthofenia* respectively.) Arrows indicate positions of apices of valves; circles, positions of hinge-axes; thicker lines, interareas.[86]

Nevertheless, some of the more stable amino-acids have been detected in shells of various ages back to the Ordovician (*c.* 400 million years).[53]

The categories of chitinophosphatic and calcareous do not quite coincide with the division between the inarticulates and the articulates; a minority of inarticulates (e.g. the living *Crania*) have calcareous shells which in composition resemble those of articulates. The chitinophosphatic type is likely to be the more primitive, but the geological evidence on this point is equivocal, since calcareous articulates (billingsellacean orthides) and inarticulates (obolellides and kutorginides) occur in Lower Cambrian strata along with some of the earliest known chitinophosphatic shells.

It is clear that the two types of composition are more or less equivalent from a functional point of view, as means of achieving a robust protective shell. But they are unlikely to be equivalent biochemically. It appears to be easier for an organism to extract the relatively rare phosphate from sea-water, and to precipitate it externally to form a shell, than to do the same with the far more abundant carbonate. In the latter case more elaborate biochemical mechanisms seem to be required, in order to remove the complex phosphates which would otherwise inhibit the growth of the crystal lattice.[91] Yet once this biochemical 'trick' had been acquired, it would open up a far more abundant source of skeletal material.

This interpretation seems to be supported, at least in broad outline, by the fossil record of brachiopods (Fig. 15). Although calcareous shells are found among the earliest known fossil brachiopods, they remained in a minority for the whole of the long Cambrian period. It is only in the Ordovician period that the great diversification of the articulates swung the balance, and put calcareous shells well into the majority. Perhaps significantly, two very different groups of inarticulates (trimerellaceans, craniaceans) appeared with calcareous shells at about the same time, and one of them seems to have precipitated aragonite instead of calcite.[52] Neither of these groups is likely to be related to the early Cambrian groups already mentioned, and one of those in turn (obolellides) is very different from the earliest articulates: thus there is evidence that the calcareous type of shell may have been evolved at least four times, three times calcitic and once aragonitic. After the Ordovician period, chitinophosphatic shells never formed more than a small minority of brachiopod faunas, although they survive to the present day.

Structure of the mantle

In living brachiopods the mantle is a very thin sheet of tissue lining the inner surface of each valve. The mantle must have been equally thin in many fossil brachiopods, for the space between the peripheral parts of the valves is often extremely narrow (Fig. 24). The mantle of living articulates consists of little more than two layers of epithelial cells, an outer layer adherent to the shell and an inner layer lining the mantle cavity, separated by a thin central layer of connective tissue. Posteriorly, where the 'body' is situated, the inner epithelium rises away from the valve surfaces and forms the outer layer of the equally thin *body wall*. The inner layer of the body wall is another sheet of epithelial cells, which lines the entire coelom. In some articulates, *spicules* of calcite are secreted by cells in the connective tissue of the mantle and body wall. Their function is unknown. They are found occasionally in fossil brachiopods as far back as the Cretaceous period,[96] and they have been found as dispersed micro-fossils even in early Jurassic strata.[68]

At the mantle edges two lobes are differentiated, comparable to the trilobed mantle edges of bivalve molluscs but generally much less conspicuous. The outer lobe is adherent to the edge of the valve; the inner lobe is often sensory in function (Fig. 9). The growth of the mantle occurs almost exclusively by proliferation of cells in a narrow

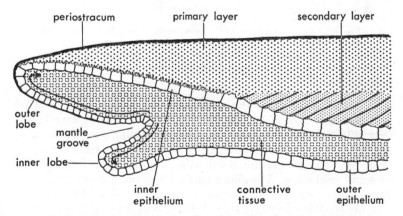

Fig. 9. Stylised longitudinal section of the edge of a valve of a living rhynchonellide (*Notosaria*), to show the structure and mode of growth of the mantle and the shell.[105] Note the 'conveyor-belt' system by which the cells formed in the generative zone within the mantle groove shift in relative position and change in secretory activity.

generative zone within the *mantle groove* between the two lobes. From here, the new cells migrate around the tip of each lobe like the treads of an advancing tracked vehicle, and so join the outer or inner epithelium.[105] Thus the growth of the mantle is largely confined to its periphery.

This may explain the characteristic form of the chief circulatory system of brachiopods. The main coelom penetrates the connective-tissue layer of the mantle as a series of branching *mantle canals* (='pallial sinuses'). Characteristically, and especially in the articu-

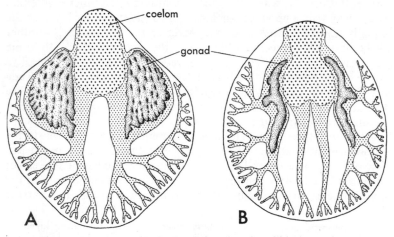

Fig. 10. Mantle canals of (A) a living rhynchonellide (*Hemithyris*: ventral mantle) and (B) a living terebratulide (*Macandrevia*: dorsal mantle), to show repeated branching towards mantle edge and regularly spaced terminal branches; note also position of gonads.[49,86]

lates, these canals leave the coelom as one or two pairs of trunk vessels in each mantle, and divide repeatedly towards the mantle edge (Fig. 10). The fine terminal branches end blindly at fairly equal intervals all round the mantle edge, very near the generative zone (Fig. 51). The canals are lined with evaginations of the ciliated coelomic epithelium. The cilia create a crude two-way circulation by drawing the coelomic fluid outwards along one side of the canal and returning it down the other side.[117] There is also a separate system of fine 'blood vessels' lying within the mantle canals: this system is apparently open-ended, but even its anatomy is poorly known and its physiology not at all. But in any case, circulation within the canals is probably responsible for conducting metabolites to and from the generative zone. Impressions of the mantle canals are often preserved on the inner surfaces of the valves of fossil brachiopods, owing to slight differences in the rate of thickening of the shell on the sites of the canals. Where the pattern is clear, they generally show the same highly divided branches terminating very close to the valve edges (Fig. 53).[109]

No critical physiological work has yet been done on the mantle canal system. But in addition to the function already mentioned, the canals almost certainly aid in respiration. There is no special respiratory organ in brachiopods, but the large surface area of the mantle and body wall is probably the main site of gaseous exchange. In one living inarticulate (*Glottidia*) the mantle surface is extended into thin-walled ampullae projecting into the mantle cavity.

Structure of the shell[105]

In chitinophosphatic shells the organic material is intimately mixed with the phosphate (e.g. *Discinisca*) or else forms alternating layers (e.g. *Lingula*). There may have been much greater variety of structure among the early Palaeozoic inarticulates, but their preservation is rarely good enough for detailed study. Chitinophosphatic shells are never penetrated except by the finest cytoplasmic threads: these extremely fine perforations have been seen in *Lingula* and related brachiopods as far back as the Silurian period.

In calcareous inarticulates (e.g. *Crania*) the calcite occurs in the form of irregular fibres, but the distinctive micro-structure of articulate shells is not found: this supports the inference that they represent independent developments of the calcareous type of shell. In both chitinophosphatic and calcareous shells the outer surface is covered by a very thin sheet or network of organic material, the *periostracum*. It is not preserved in fossil brachiopods.

In articulates the calcareous shell usually has two or three distinct layers (Fig. 9); and well preserved fossil shells suggest that this arrangement has existed since very early in the history of the class: it is found, for example, in Cambrian orthides which probably lie near the root stock of all later articulates.[106]

The very thin outer *primary layer* is generally composed of extremely fine granular calcite, which is secreted by the mantle cells only at the extreme edge of the mantle lobe. The primary layer is therefore confined to the outer surface of each valve. In all the strophomenides its micro-structure is abnormal, and it consists of thin, neatly stacked laminae of calcite. The inner *secondary layer* (or 'fibrous layer') is composed of slender calcite fibres, stacked in a characteristic pattern, and inclined at a low angle to the surface of the shell. The fibres are separated by thin strips of protein. This layer is secreted by the whole surface of the mantle epithelium within the extreme marginal zone. Each cell secretes a single calcite fibre with its associated strip of protein. This secretion generally continues throughout life, so that the secondary layer increases in thickness away from the valve edge. Very fine growth-lines on the fibres may indicate diurnal deposition. In most of the strophomenides the standard stacking pattern of the fibres is almost completely obscured by a secondary breakdown of the fibres into bands of calcite and protein, producing a highly characteristic micro-structure; but growth-lines, possibly diurnal in origin, are still visible. In some brachiopods (especially the spiriferides) a third *prismatic layer* is laid down within the secondary layer; it is composed of much coarser fibres oriented perpendicular to the surface, and it is formed by the breakdown of the boundaries between the secondary-layer fibres, due to the cessation of protein secretion.

Although the histochemistry involved is still poorly known, this layered structure clearly reflects successive changes in the secretory activity of the mantle cells. While a newly formed epithelial cell is still on the inner side of the outer lobe, it secretes periostracum. With further outward growth of the mantle edge it migrates round the tip of the outer lobe. As it reaches the outer surface it begins to secrete primary layer. When further growth displaces it into a position relatively further from the edge, it changes quite abruptly to secreting a single secondary layer fibre, with its associated protein sheet on the outer side. With continued secretion, and a gradual shift in position, a long oblique fibre is formed. Later still, if protein secretion ceases, the boundaries of the fibre wall break down, and the cell merely contributes towards one of the coarse prisms of the prismatic layer. Thus the primary layer determines the external shape of the shell, and its irregularities in deposition form the growth-lines; the secondary layer increases the thickness and strength of the valves internally; and the prismatic layer, if present, provides further thickening and weight, often in the posterior part of the shell.

The micro-structure of the primary and secondary layers, and their presence in some of the earliest articulates, strongly suggest that they are homologous in all articulates. It is therefore possible to infer the extent of the mantle tissue in any extinct species from the distribution of the layers. The edges of the area covered by the primary layer represent the edges of the mantle; the areas on which the secondary layer is exposed represent the areas formerly covered by the mantle.

The application of this principle confirms that in most fossil brachio-pods, as in all living species, the mantle was confined to the internal surfaces of the valves. In a few strophomenides, already mentioned as having lost the protective function of the shell, the dorsal valve seems to have become entirely sheathed in mantle tissue (Fig. 5B, C). Its outer surface generally has no growth-lines, and resembles the adjacent *inner* surface of the ventral valve; and the secondary-layer material of which it is composed seems to have been thickened by accretion on both surfaces. This reconstruction accentuates the aberrant nature of these brachiopods.

Many articulates have shells with the layered structure described above, and without any perforations. Such brachiopods are termed *impunctate*. There are two other types of shell structure among the articulates. *Punctate* shells are perforated with a series of fine holes: they are discussed further in the next section of this chapter. *Pseudo-punctate* shells, as the term suggests, often appear to be punctate, but detailed study of thin sections shows that they are not. In some pseudopunctate shells the fibres of the secondary layer are merely warped inwardly in a conical fashion around a series of irregularly spaced points, which thus emerge on the internal surface of the shell as a series of small tubercles.[109] With further internal thickening, 'centres' of warping extend more or less obliquely right through the shell. This may have been the original form of pseudopunctation, and differs from the impunctate structure merely in the warping of the fibres and consequent pustular internal surfaces of the shell. It is widely distributed among strophomenides, but probably developed at least twice within that order; it is also known in a small group of orthides (gonambonitids) and also in a few terebratulides. In some strophomenides a more complex type of pseudopunctation is found. Along the axis of each pseudopuncta is a slender rod or *taleola* of finely granular calcite (Fig. 11). This was evidently due to differential

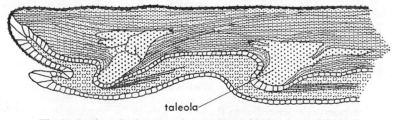

taleola

Fig. 11. Section of valve edge of a strophomenide (*Leptaena*, Silurian) to show taleolae; mantle and periostracum reconstructed by analogy with living brachiopods.[109]

secretion by epithelial cells on the tip of each pustule. The material of taleolae is similar to the granular calcite often secreted on the sites of attachment of the muscles: it is therefore possible that taleolae represent points of attachment of muscle fibres that gave the mantle tissue some degree of contractility. From an evolutionary point of view taleolae are perhaps less important than the aberrant micro-structure mentioned above, which appears to be almost uniquely diagnostic of the Strophomenida.

Mantle caecae[61]

In punctate brachiopods the shell is perforated by large numbers of fine cylindrical holes or *punctae*, which in the living state contain slender extensions or *caecae* of outer mantle epithelium. Punctae and caecae extend perpendicularly through almost the entire thickness of the shell (Fig. 12). They are present even in the extreme peripheral zone of each valve, and are evidently initiated there; later, as new

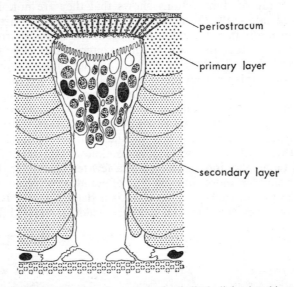

Fig. 12. Stylised section of a caeca in the shell of a living brachiopod, showing its blunt termination close to the external surface of the shell. This 'head' is largely filled with secretory cells exuding mucoprotein, which may seep outwards through the 'brush' tubules to the periostracum. The hollow 'stalk' of the caeca lengthens as the shell thickens but remains in contact with the mantle epithelium.[61] Nuclei black, mucoprotein inclusions irregularly stippled. Compare with Fig. 9.

shell material is secreted on the inner side of each valve, the caccae are prolonged through the new layers so that they remain in contact with the rest of the mantle tissue. Caecae invariably terminate just short of the external surface of the shell, though a 'brush' of extremely fine cytoplasmic threads extends from the blunt 'head' of the

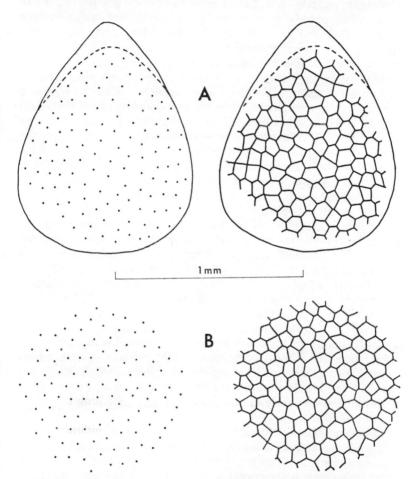

Fig. 13. Pattern of punctae on (A) a small juvenile terebratulide shell (sp. indet.), and (B) a small part of an adult terebratulide (*Magellania*). In each case the right-hand figure has been converted into a series of 'cells', each centred on a puncta, to illustrate the approximation to hexagonal close-packing.[32]

caecae through the very thin remaining part of the primary layer. Caecae are formed on every part of the valve edges throughout ontogeny, and are therefore found all over the valve surfaces. Their outer ends are distributed uniformly over the surface: in some groups they show quite accurate hexagonal close-packing (Fig. 13).[32] Generally punctae can be detected without difficulty in fossil material, though the preservation is sometimes ambiguous and the study of thin sections is always advisable. They are, however, destroyed by extensive recrystallisation or replacement.

The exact function of caecae is uncertain. They seem unlikely to be sensory, for they contain no apparent receptors, no pigment spots and no nerve endings. Yet their termination as a series of regularly spaced points immediately below the outer surface of the shell does suggest some function directed to the interface between shell and environment. This possibility is strengthened by the fact that the 'head' of each caeca is filled with special secretory cells, which appear to exude mucopolysaccharide into the space immediately beneath the 'brush'. From here it may seep outwards through the tubules of the brush, but its exact function on reaching the periostracum is still not entirely clear.

No true intermediate is yet known between the impunctate and punctate conditions. A few Palaeozoic brachiopods have been termed 'exopunctate' (the punctate condition being sometimes termed 'endopunctate') in the belief that a pattern of shallow pits on the external surface of the shell might represent the sites of transient caecae which were active only at the valve edges and later blocked off. But in no case has this interpretation yet been substantiated convincingly by thin sections, and in some the alleged exopunctae are almost certainly the broken bases of small spines. On present knowledge, therefore, it must be concluded that punctation is an all-or-none character. Since punctae are also invariably distributed all over the shell, and even at the apices of the valves, the structure must have been initiated very early in ontogeny, and may have evolved paedomorphically.

The taxonomic distribution of the punctate structure is curiously scattered, and cuts right across the accepted high-level classification. Among the predominantly impunctate orthides there is one large punctate group (enteletaceans). Similarly there is one punctate group (retziaceans) among the mainly impunctate atrypides. The spiriferides are also predominantly impunctate, but include at least two distinct punctate groups (suessiaceans, spiriferinaceans). The pentamerides are all impunctate, and the terebratulides all punctate; but a few isolated punctate genera have been reported in both the rhynchonellides and the strophomenides.

This distribution suggests a multiple evolutionary origin for the punctate structure, and this conclusion now seems unavoidable. However, the number of times that this occurrence must be postulated is not as great as might at first appear, since several of the punctate groups may be related to one another (Fig. 15). Moreover, some alleged cases of punctation have been discredited after more critical study. Nevertheless, it is a striking fact that punctate and impunctate groups often seem to have evolved on remarkably parallel lines, or to have converged remarkably closely. For example, there is a punctate orthide (*Pionodema*) which occurs in the same Ordovician strata as an impunctate one (*Mimella*), and which is so similar externally that until the significance of punctation was realised both were assigned to the same species.[27] Yet such cases of *homoeomorphy* are certainly not due, for example, to sexual dimorphism, for closer study of the whole morphology shows that their affinities clearly lie with quite different groups. The same phenomenon occurs also on a higher taxonomic level. For example, each of the three groups of punctate atrypides and spiriferides mentioned above seems—on all other criteria—to be more closely related to a different impunctate group than to any other punctate group. The position is further complicated by the suspicion that the process was reversible, and that some punctate groups may have given rise to impunctate groups.

Altogether, the discovery of the scattered distribution of punctation had an extremely unsettling influence on the already tangled systematics of brachiopods. There was for a time a tendency to regard it as a panacea for the problems of systematics, and to assume uncritically that it could be given taxonomic 'weight' above all other characters. For example, one small extant group of punctate brachiopods (thecideaceans) is commonly, if tentatively, linked to the punctate terebratulides or spiriferides,[105] although in most other features it has a much closer resemblance to the pseudopunctate strophomenides.[84] However, in spite of all this confusion, a closer study of the distribution of punctation is gradually yielding a clearer picture of its probable evolutionary history (Fig. 15).[108]

Almost all Cambrian articulates are impunctate, and this was probably the primitive condition of the articulate shell. Punctation first appeared for certain among the orthides in the early Ordovician period. It is present in three rather distinct families, which may have evolved the structure independently,[111] but are perhaps more probably related to each other. They seem to have given rise to all the later punctate orthides. Punctate spiriferides (suessiaceans) and punctate atrypides (retziaceans) both first appeared at about the same time in the Silurian period, but do not seem closely related. Early in the Carboniferous period, a second group of punctate spiriferides (spiriferinaceans) made their appearance, but do not seem closely related to the first. All these punctate groups were quite abundant throughout the later Palaeozoic, and most of them survived to the end of the Triassic period or even into the early Jurassic. Their impunctate relatives were equally important up to the end of the Palaeozoic era, but few survived into the Mesozoic.

The exclusively punctate terebratulides first appeared in the early Devonian period, but were probably derived from some already punctate atrypide (retziacean) ancestors. Although they were often abundant even in the later Palaeozoic era, it was in the Mesozoic that they emerged as the dominant group of brachiopods, and they have remained so to the present day.

Among the rhynchonellides, punctation is only known in one very small group (rhynchoporids), which first appeared early in the Carboniferous period and survived till the end of the Palaeozoic era. It would be tempting to derive this group from some already punctate ancestors, but in all other characters its affinities are clearly with the impunctate 'mainstream' rhynchonellides (rhynchonellaceans), and it almost certainly represents yet another independent development of punctation. Finally, punctation occurs among the rare post-Palaeozoic survivors of one strophomenide group (davidsoniaceans), and one small extant group (thecideaceans) is probably related to these.

This synthesis of the history of punctation will almost certainly prove to be incorrect in several respects, as further study reveals the distribution of the structure with greater precision. But two points which emerge from it are unlikely to be seriously modified.

First, punctation must have been evolved several times independently. It is fair to add, however, that closer study of the micro-anatomy of punctae may show greater diversity in their structure than has hitherto been assumed; and this may allow greater confidence in assigning punctate groups to separate points of origin. Already, for example, punctae which are not blunt-ended but have many tapering branches have been reported in one orthide[112] and one strophomenide.[84] These are not unlike the punctae of some inarticulates (craniaceans) (Fig. 14), which have never seriously been considered as close relatives of any of the articulates, and undeniably represent an altogether separate development of punctation.

The second point that emerges from the history of punctation is the curiously uneven 'success' of the punctate groups, relative to the similar impunctate groups. Among the orthides, the impunctate became extinct in the late Devonian period, whereas the punctate survived 100 million years longer till the end of the Permian. But there is little difference in the overall time-range or relative abundance and diversity of the impunctate and punctate atrypides and spiriferides, while the only development of punctation among the

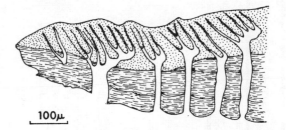

100μ

Fig. 14. Section of edge of valve of the living acrotretide *Crania*, to show branched punctae extending towards outer surface;[13] mantle tissue and caecae omitted.

rhynchonellides was singularly unsuccessful relative to the rest of this abundant group. The exclusively punctate terebratulides have indeed been the dominant brachiopods since early in the Mesozoic era, but it is not clear that this is due even in part to their punctation. Although the impunctate rhynchonellides are now less common and diverse, they can be found living in the same habitats as terebratulides, and are at no obvious disadvantage.[76] However, at one such locality there is a significant difference in the incidence of shell-boring organisms, which penetrate the impunctate shells more commonly than the punctate: this suggests that the caecal secretions may inhibit such organisms.[61]

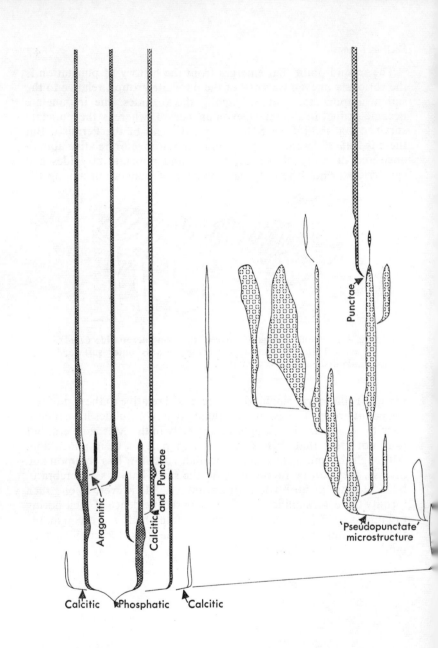

Aragonitic

Calcitic and Punctae

Calcitic   Phosphatic   Calcitic

Punctae

'Pseudopunctate'
microstructure

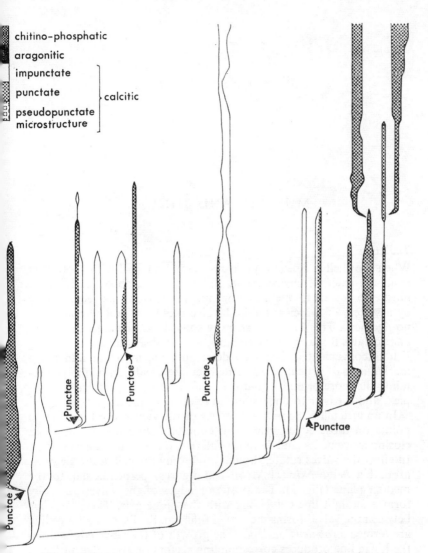

chitino-phosphatic

aragonitic

impunctate

punctate ⎫ calcitic

pseudopunctate
microstructure ⎭

Punctae

Punctae

Punctae

Punctae

Punctae

Punctae

Fig. 15. Chart to show possible evolution of shell composition and structure in the Brachiopoda. Each 'spindle' represents a super-family, and is derived from the data used for Fig. 99 by drawing a smoothed 'envelope' around the 'bundle' of lines representing the ranges and generic diversities of the constituent families. Thus the width of each spindle at a given point in geological time is a rough indication of the relative diversity of that superfamily at that time, estimated by a combination of familial and generic levels. Super-families may be identified by comparison with Fig. 99. Solid arrow-heads indicate possible points of evolution of 'key' features.[86,105,108]

3

MUSCLES AND HINGES

The hinge

When the shell is closed it provides the brachiopod with protection from the external environment. But for all metabolic activities it must also be able to open. The opening and closing is effected solely by a system of muscles: unlike bivalve molluscs, brachiopods have no ligament. There is also generally some kind of hinge, though in inarticulates it may be very rudimentary or absent altogether.

It is convenient to begin by describing the muscular and hinge mechanisms of articulates. The inarticulate condition is almost certainly the more primitive, but it is also in some respects more complex, and is easier to understand in relation to the articulate.

In an articulate the posterior edges of the valves, or at least some points on those edges, remain in contact during the opening and closing movements, and form the fulcrum for the muscular leverage. In effect the valves rotate, though often only through a few degrees, around a *hinge axis* (H.A. on text figures) perpendicular to the median plane (Fig. 2). The posterior edges of the valves generally form a straight line coinciding with the hinge axis. This *hinge line* between the valves forms the actual fulcrum. Shells with a hinge line are termed *strophic* (Fig. 16A). The growth of the valve edges along the hinge line produces corresponding sectors on the valve surfaces. These are roughly triangular *interareas* (or 'cardinal areas'); each has its apex at the apex of the valve, its base is the hinge line, and its growth-lines mark the former positions of the hinge line during the growth of the shell. The rate of growth is generally greater on the ventral side, so that the ventral interarea is generally a higher isosceles triangle than the dorsal (Fig. 8B, E, F). If either valve grows into a high conical shape, the interarea is extended similarly (Fig. 8H).

Conversely, with a low rate of growth one or both interareas may become vanishingly low (Fig. 8G). The shells of some articulates have no hinge line, and the fulcrum is reduced to a pair of points on the posterior valve edges; these shells are termed *non-strophic* (Fig. 16B). Although non-strophic shells predominate among living articulates, all the earliest articulates were strophic; there is no doubt that a hinge line was an original character of the articulates, which in some groups has since been lost. If the hinge line is lost, interareas are of course lost also.

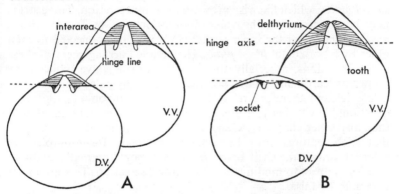

Fig. 16. Diagrammatic views of strophic (A) and non-strophic (B) shells, to show how interareas of strophic shells are formed by accretion on a hinge line coinciding with the hinge-axis; in the non-strophic shell no portion of the growing edges (thick lines) lies along the hinge-axis, and there are therefore no true interareas.[71]

Articulation

The contact between the valves at the fulcrum is generally strengthened by an internal articulation, which rigidly confines the movements of the valves to rotation around the hinge axis, and which prevents lateral slewing. The primary articulation consists of interlocking *teeth* and *sockets*. These lie just inside the posterior edges of the valves and just in front of the hinge axis (Fig. 16), so that when the shell opens or closes they slide against one another (Figs. 20, 21). In living articulates the teeth and sockets interlock so closely that the valves cannot be forced open, beyond the natural angle of gape, without breaking the articulation altogether. Consequently the valves generally remain together after death. Most post-Palaeozoic brachiopods shared this characteristic, and are usually preserved in the fossil state with both valves together. But in many earlier groups the interlocking seems to have been relatively loose, and the valves were easily separated after death (like those of most bivalves).

The ventral valve has a single pair of teeth; the dorsal valve has a corresponding pair of sockets (Figs. 16, 19, 22). The teeth are bounded on the median or anterior side by the inner socket walls, which are often as large and prominent as the teeth themselves. Teeth and socket walls are composed of secondary-layer material, and grow by accretion during ontogeny. They may be connected to the floor of the valves by reinforcing struts, termed *dental plates* and *socket plates*, which likewise grow forward by accretion. These struts may be connected to the valve floor indirectly, by fusion with a *median septum*; or the inner socket walls may be connected to each other by a horizontal *hinge plate*. The detailed arrangement of these shelly partitions, especially those in the dorsal valve (collectively known as 'cardinalia'), have been widely used by palaeontologists as a 'stable' and therefore reliable character in taxonomy. But it is questionable whether it is any less subject to convergent adaptation than any other character. Generally there is an implicit assumption that such features must be more stable than, for example, the external form of the shell, because they are more 'internal' and therefore less subject to modification by natural selection. This argument is manifestly false.

On the other hand, the constancy of the dentition itself is probably a reliable sign that articulation was only evolved once, and that the articulates are therefore a natural monophyletic group. For this particular arrangement, of a pair of teeth and a pair of sockets in a constant relation to each other, is only one of many possible types of dentition, all of which could be functionally equivalent. In this respect, the articulate brachiopods form a striking contrast to the bivalve molluscs, among which the dentition is extremely diverse.

A few extinct articulates modified or lost their original articulation, but there is never any doubt that these are secondary modifications, for the associated hinge and musculature remain 'articulate' in character. In some spiriferides the normal articulation was supplemented by the development of interlocking *denticles* all along the hinge line. Denticles also evolved independently in several strophomenides (e.g. stropheodontids). But in the latter, a gradual spread of denticles outwards along the hinge line was accompanied by a progressive reduction and final atrophy of the original teeth and sockets (Fig. 17). This suggests that the denticulation represents a functional replacement. Its precise adaptive significance is not yet clear, but it might have been related to the development of a feeding

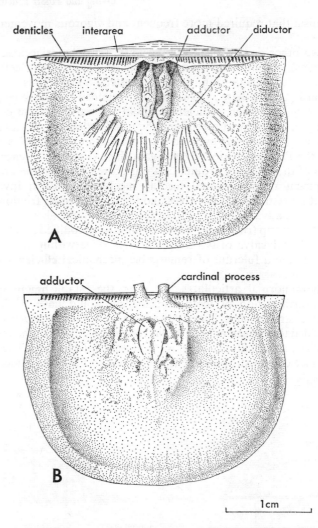

Fig. 17. Interiors of ventral (A) and dorsal (B) valves of a Devonian strophomenide (*Stropheodonta*; Strophomenacea), with normal articulation replaced by denticles along hinge line (from photographs in [103]).

mechanism that required more frequent and vigorous movements of the hinge.

In two other groups of strophomenides (strophalosiaceans, productaceans), all but the earliest Devonian genera lost their teeth and sockets without acquiring denticles instead. They thus became secondarily 'inarticulate'. However, they retained broad hinge lines, which may have been sufficient to hold the valves in the correct orientation, and they also retained the usual articulate musculature. One genus (*Ctenalosia*) evolved an articulation of denticles later, during the Permian period. Here again there is some evidence that the loss of the ordinary articulation was correlated with the gradual development of an unusual feeding mechanism, which involved frequent movements of the valves and for which a more friction-free hinge was an advantage (p. 145). In the most extreme aberrant members of this group (richthofeniids) the hinge was modified still further and the dorsal valve rests on a knife-edge ridge within the ventral valve, forming a fulcrum of remarkable mechanical efficiency (Fig. 18).

In more normal articulates, however, the establishment of an effective articulation of teeth and sockets would have made the hinge line strictly superfluous: the teeth and sockets alone could have guided the movements of the valves and held them in the correct orientation, especially if they were closely interlocked. The first appearance of non-strophic shells, which had lost the hinge line, may

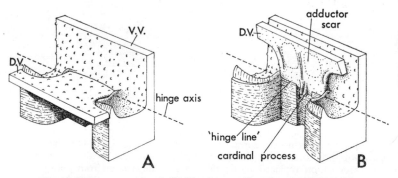

Fig. 18. Hinge structure of a highly aberrant Permian strophomenide (*Prorichthofenia*; Strophalosiacea), with internal trapdoor-like dorsal valve closed (A) and open (B): block-diagrams of hinge region (cut surfaces of shell blank). The dorsal valve is balanced on a 'hinge line' consisting of a knife-edge ridge; there is also a fine axle-like lateral bearing.[73]

therefore mark the achievement of an articulation rigid enough to stand by itself. Probably the hinge line became narrower and narrower, until it (and the interareas too) disappeared. The first non-strophic shells are found in Ordovician strata, among the earliest rhynchonellides, and thereafter a non-strophic hinge is characteristic of that group (Fig. 33). It was probably inherited by the atrypides from rhynchonellide ancestors, and passed on by them to the terebratulides. Some of the later pentamerides also have non-strophic hinges, which were probably evolved independently. Conversely there is no inherent reason why a strophic hinge should not have been re-evolved, simply by a slight change in the orientation of the posterior valve edges. There is good evidence that this did in fact happen more than once: in both Devonian and Triassic strata there are related species with all gradations of structure between the clearly strophic and the clearly non-strophic (*Anathyris, Clavigera*); and the sporadic appearance of strophic shells among the pre-dominantly non-strophic terebratulides suggests the same process (e.g. stringocephalids, megathirids, platidiids).

Interlocking structures

Although the internal articulation of teeth or denticles, and indeed even a hinge line, would have served to guide and confine the movements of the valves, many articulates acquired further articulation of a rather different kind around the rest of the valve edges. Various *interlocking* structures ensured an accurate fit between the valve edges during the final phase of the closing of the shell. In the simplest form, the valve edges were deflected into complementary crenulations, so that the 'crests' of one valve edge fitted into the 'troughs' of the other (Fig. 55A). In the course of ontogeny this produced a series of corresponding radial ridges or *costae* radiating from the umbo across the surface of each valve. In many shells the interlocking of the extreme edges of the valves was supplemented by a series of interlocking buttresses of secondary-layer a little further in (Fig. 19): these are very closely paralleled in some bivalve molluscs (e.g. *Cardium*). Interlocking structures first became abundant among the Ordovician orthides, and remained common in many groups in later periods. But only a few of them survive (e.g. *Terebratulina*). This may be due to an inverse functional correlation with the normal dentition: for in one lineage of Silurian atrypides (*Eocoelia*) a gradual strengthening of the dentition coincided with a weakening of the costae,[114] and therefore of the interlocking. The exact functional significance of strong dentition and/or interlocking structures is not altogether clear:

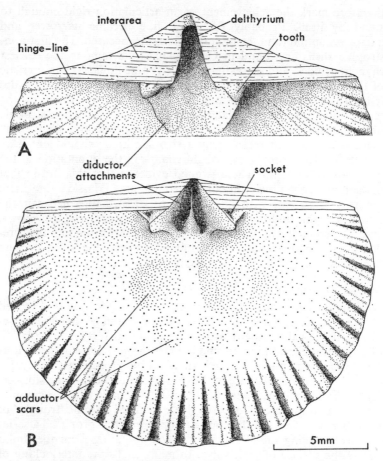

Fig. 19. Interior of *(a)* ventral and *(b)* dorsal valves of an Ordovician orthide (*Hesperorthis,* Orthacea), to show hinge line, teeth and sockets, and marginal interlocking buttresses.[86]

possibly they afford protection against certain predators, such as asteroids.[20]

Muscular leverage

In all articulates one pair of muscles is clearly responsible for closing the shell: the *adductor* muscles extend more or less vertically through the coelom near the anterior body wall, and their line of action is well in front of the hinge axis (Fig. 20). They are strongly attached to the inner surface of the valves, the intervening epithelium being modified by the development of tonofibrils which even penetrate the secondary-layer fibres. The adductors are differentiated into two portions, showing a remarkable similarity to the adductors of bivalve molluscs.[74] The smaller posterior adductors consist of 'quick' striated muscle fibres, which snap the shell shut in response to various sensory stimuli. The larger anterior adductors consist of unstriated 'catch' fibres, which react more slowly and hold the shell tightly closed for long periods. The posterior and anterior adductors

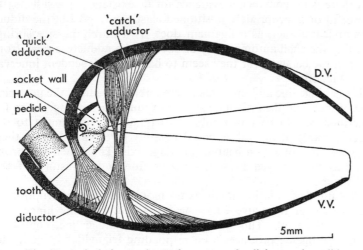

Fig. 20. Articulation and muscle system of a living terebratulide (*Waltonia*; Terebratellacea), in median section: note their relation to the hinge-axis.[74]

form separate bundles of fibres on the dorsal side, but they share a single attachment to the ventral valve (Figs. 20, 28). Evidence of the same differentiation can be seen in many fossil articulates, where there are two pairs of adductor *muscle scars* on the inner surface of the dorsal valve, and a small median scar or pair of scars on the ventral valve. The preservation of muscles scars, which are slightly raised or depressed areas of valve surface, is due to slight differences in the rate of secretion of the secondary layer at the sites of attachment, or to a modification of the microstructure of the shell.

The shell is opened by the contraction of a second pair of muscles. The *diductor* (or 'divaricator') muscles run obliquely across the coelom, from the posterior tip of the dorsal valve on to the floor of the ventral valve; their line of action invariably passes the hinge axis on the opposite side from the adductors. Their contraction draws the posterior edges of the valves closer together, and so causes the front of the shell to gape open (Fig. 20). The adductors and diductors thus act antagonistically on opposite sides of the hinge axis to close or open the shell, but the moment of the diductors is always much smaller than that of the adductors. In brachiopods there is no ligament like that of bivalve molluscs. (A periostracal 'pad' with the function of a ligament has been inferred in some fossil brachiopods,[92] but there is no convincing evidence for its existence as a sufficiently powerful or appropriately positioned elastic spring.) But the diductors appear to act, as a ligament does, as a purely passive spring, opening the shell automatically as soon as the adductors relax from the contracted state, for they seem to have no independent innervation (Fig. 50).

If the hinge line and interareas were unbroken across the posterior side of the shell, the articulate system of muscular leverage, as so far described, would be unworkable. No points of attachment to the inner surfaces of the valves could possibly give the diductors their necessary line of action behind the hinge axis. This functional problem has been solved in several ways during the history of the articulates.

The simplest solution involves no more than the interruption of the growth at the valve edges near the centre of the hinge line, between the teeth. This median gap in the hinge line, continued during ontogeny, produces a corresponding triangular space in the centre of each interarea. These spaces may be termed *delthyria* (Figs. 16, 19). (The dorsal delthyrium has generally been termed a 'notothyrium': this is only one instance of the widespread duplication of descriptive terms for brachiopods; when the structures concerned are clearly homologous on the two valves, this duplication serves no useful purpose and only creates confusion.) Together the delthyria form a kite-shaped hole, the *delthyrial gap*, on the posterior side of the shell. This is characteristic of most of the orthides (orthaceans, enteletaceans) and many of the spiriferides. As long as the umbonal regions of the valves project posteriorly from the hinge line, which in most brachiopods they do, this gap would have enabled the diductors to pass from one valve to the other on an oblique line behind the hinge axis (Fig. 21). In doing so, the muscles would necessarily have passed outside the shell and must have been covered by the posterior body wall. In principle this would mar the protective efficiency of the shell; in practice the delthyrial gap was generally held closely against the substratum, for in many orthides it was also used as a point of emergence for the pedicle. But its primary function was not as a pedicle foramen but as a means of securing leverage for the diductors.

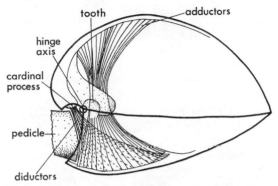

Fig. 21. Articulation, and reconstruction of muscle system, of a simple Ordovician orthide (*Hebertella,* Orthacea); lateral view, drawn as though shell were transparent. Note diductors passing outside shell on posterior side, and antagonistic adductors.[34]

Another solution to the problem allows the diductors exactly the same line of action, but gives them greater protection. The valve edges, instead of being interrupted in the centre of the hinge line, are merely deflected posteriorly (Fig. 22). Of course they cannot remain precisely in contact, or they would prevent the valve edges in front of the hinge axis from gaping apart. But only a very narrow gap —presumably covered by the body wall—has to be left between them. These deflected parts of the valve edges, continued during ontogeny, produce a corresponding convex triangular area in the centre of each interarea. Under these convex vault-like *delthyrial covers** the diductors could pass from one valve to another on a line

---

* They are commonly described as though they were separate plates like the plates of an echinoderm skeleton; in fact they are merely modified parts of the ordinary valve surface. A bewildering and largely superfluous proliferation of morphological terms has obscured the essential simplicity of delthyrial covers.[34] The dorsal cover (of the 'notothyrium') is termed a 'chilidium'. The ventral cover is a 'deltidium' or 'symphytium' if the presence of a pedicle in the gap has forced it to form initially as two separate parts ('deltidial plates'), growing inwards across the gap and later fusing into a single cover (Fig. 38). If no pedicle emerges through the delthyrial gap the growth of the cover is not interrupted in the same way, and the cover is then termed a 'pseudodeltidium'. The difference is important for the phylogeny of pedicle attachment (see p. 80), but the delthyrial covers themselves are identical in function and often similar in appearance.

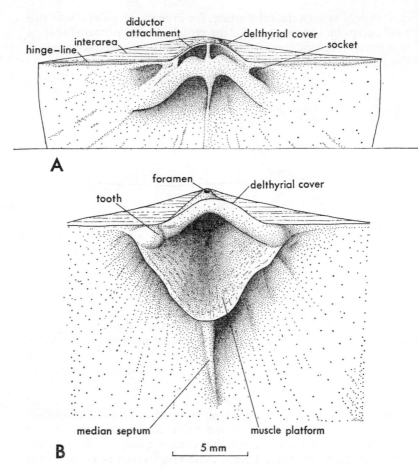

Fig. 22. Articulation in an Ordovician orthide (*Estlandia*; Clitambonitacea). Interiors of posterior parts of both valves, showing teeth and sockets, delthyrial covers, and muscle platform.[86]

behind the hinge axis (Fig. 23). Convex delthyrial covers are found in the earliest orthides (billingsellaceans) and in some later groups (clitambonitaceans). They are also characteristic of most of the strophomenides and of some spiriferides (e.g. suessiaceans). Some other spiriferides (some spiriferaceans and spiriferinaceans) developed functionally equivalent covers in the form of *stegidial plates*. Unlike the many other so-called 'plates' in brachiopods, these were entirely separate from either valve, and grew by peripheral accretion to fill the enlarging delthyrial gap.[33]

A third solution to the problem of leverage may develop out of either of the first two. If growth on the posterior side of the dorsal valve is sufficiently slow, the dorsal umbo will become highly incurved. If the hinge is non-strophic, or the hinge line is narrow enough, the dorsal umbo may then become tucked inside the ventral delthyrium. Then the diductors will be well protected within the

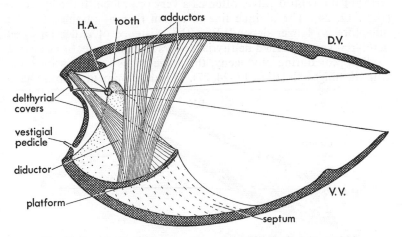

Fig. 23. Reconstruction of muscle system of an Ordovician orthide (*Estlandia*; Clitambonitacea) with delthyrial covers and raised muscle platform.[86] Compare with Fig. 22.

shell cavity, and their attachment to the dorsal umbo will give them an oblique line of action on the opposite side of the hinge axis to that of the adductors (Fig. 26). No dorsal delthyrial cover is required; that part of the ventral delthyrium which is not occupied by the dorsal umbo may have a cover of some kind, or may be largely occupied by a pedicle (Fig. 20). This solution to the problem of leverage is the one that is most widespread among living brachiopods, being characteristic of non-strophic shells. It is universal in the rhynchonellides and terebratulides, and common in the pentamerides and atrypides.

Except among the earliest articulates, the dorsal attachments of the diductors are very commonly thickened and strengthened by a pad or knob of secondary-layer material. This is termed the *cardinal process* (Fig. 21). The paired attachments can often be seen on its surface. During the evolution of the articulates it frequently became large enough to affect the line of action of the diductors. In this way it might fill most of an open dorsal delthyrium, or most of the space beneath a dorsal delthyrial cover, thus shifting the diductors into a more oblique orientation.

This development led to the fourth and last solution to the problem of leverage. In several groups of brachiopods the cardinal process became even larger, projecting beyond the hinge axis into the interior of the ventral valve, often as a very prominent forked process (Figs. 17, 24). The oblique line of action thereby attained for the diductors no longer depended on the existence of either an open delthyrial gap or arched delthyrial covers; and these features tended to disappear during phylogeny, the valve edges returning into line with the hinge axis (Figs. 17, 24, 27). This development seems to have

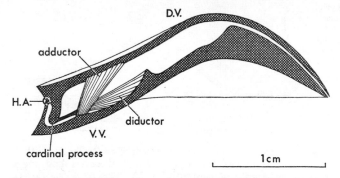

Fig. 24. Reconstruction of muscle system of a Devonian strophomenide (*Strophonelloides*; Strophomenacea) of saucer-like shellform. Note the prominent cardinal process.[86] Compare with Fig. 17.

occurred several times, for there are several groups of articulates not closely related to one another, in which the interareas extend unbroken across the median plane (e.g. stropheodontids, thecospirids, triplesiaceans). The growth of a large projecting cardinal process also made the interareas themselves strictly redundant, at least for the function of leverage; and they were virtually lost in some groups (e.g. productaceans), as already mentioned (Fig. 25).

Muscle platforms

The muscle attachments of many articulates were raised off the floor of one or both valves (Fig. 26). Structures having this effect may be termed *muscle platforms*. (As usual, the basic similarity of these

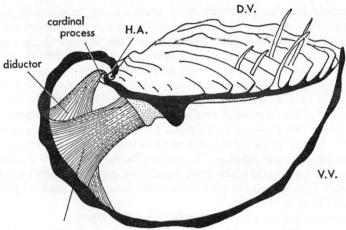

Fig. 25. Reconstruction of muscle system of a Carboniferous strophomenide (*Overtonia*; Productacea); note prominent cardinal process.[16]

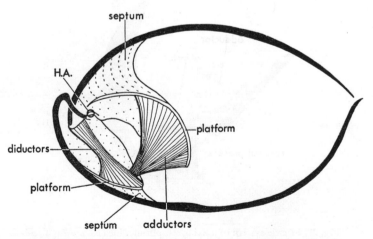

Fig. 26. Reconstruction of muscle system of a Permian rhynchonellide (*Stenoscisma*) with raised muscle platforms.[45]

structures has been masked by a proliferating terminology, e.g. spondylium simplex, spondylium duplex, pseudospondylium, crura-lium, camarophorium, shoe-lifter process, spyridium etc.: some of these terms are valid as descriptions of genuinely different varieties, but many are superfluous.) Most muscle platforms consist of some kind of shelf, composed of secondary-layer material, projecting from the back of the valve and growing forwards by accretion during ontogeny. Often the sides of the shelf are curled up, even forming a deep narrow trough ('spondylium'). Often the platform is linked to the valve floor by a median septum, which presumably improved its mechanical strength (Fig. 22). Judging by their variety of structure and form, and by their scattered occurrence among different groups of brachiopods, muscle platforms must have been evolved many times (Fig. 33).

Yet their function is not entirely clear. For any given orientation of a muscle they would have the effect of shortening its length. It may be significant that the only living brachiopod with a muscle platform (*Lacazella*) is also the only living articulate with *columnar* muscles (Fig. 27). Columnar muscles are those in which the whole length of the muscle contains contractile fibres, so that the total length of the muscle is also its effective contractile length. Most living articulates, on the other hand, have *tendonous* muscles. In

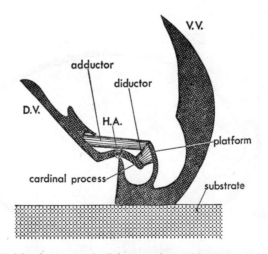

Fig. 27. Muscle system of a living strophomenide (*Lacazella*; Theci-deacea), showing 'columnar' muscles raised on a central platform. Note also cemented attachment.[57,86]

these, the contractile fibres are confined to one end of the muscle
(the dorsal end of the adductors, and ventral end of the diductors),
and the rest of the muscle consists of tendon (Figs. 20, 28); thus the
effective contractile length is much less than the total length of the

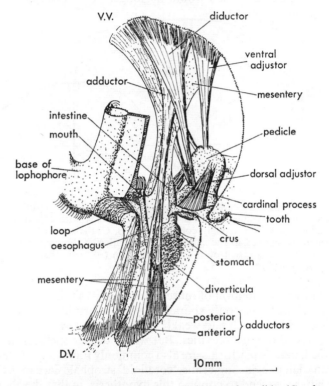

Fig. 28. Tendonous muscles in a living terebratulide (*Gryphus,*
Terebratulacea): oblique view of dissection with body wall and
near half of lophophore and loop removed. This dissection also
shows pedicle and its muscles, gut supported by mesenteries, and
base of lophophore supported on edge of brachial loop which is
embedded in body wall.[86]

muscle. For any given position of a muscle (especially its distance
from the hinge axis), the most effective contractile length may
sometimes be much less than the actual distance between the valve
surfaces. If the muscle is tendonous such a discrepancy can be made
up by a length of tendon: but if it is columnar, the muscle itself must

C

be shortened, by raising its attachments (Fig. 29). On this interpretation, muscle platforms would tend to be evolved whenever brachiopods with columnar muscles developed rather strongly convex shells. Once a group of brachiopods had acquired tendonous muscles, muscle platforms would no longer be necessary.

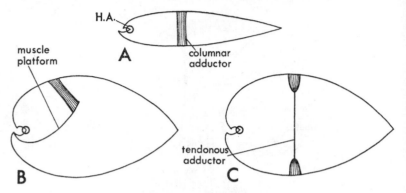

Fig. 29. Diagrams to illustrate possible function of muscle platforms and tendonous muscles. A, a gently convex shell with columnar adductors. In the strongly convex shells, B, C, the same effective length is retained in the same position relative to the hinge axis, *either* by being raised on a muscle platform (B), *or* by becoming tendonous (C).[86]

The known distribution of muscle platforms fits this interpretation quite well. They occur only sporadically in the orthides and strophomenides, most of which had weakly convex or concavo-convex shells with only shallow shell-cavities. But where deeper shell-forms were developed, muscle platforms are also commonly found (e.g. clitambonitaceans, later davidsoniaceans, 'coralloid' strophalosiaceans). Some of the latter also tend to confirm this hypothesis, in that the shortening of the muscles was achieved in quite different ways, e.g. by an exceptionally large median septum and cardinal process (Fig. 30). Platforms are almost universal in the pentamerides, which often had strongly biconvex shells, and they are also characteristic of one spiriferide group (cyrtinids). But there are only sporadic examples of muscle platforms in the equally convex atrypides, and few if any in the terebratulides. This suggests that tendonous muscles may have been evolved at some point near the origin of the rhynchonellides, and may then have been inherited by the atrypides and later by the terebratulides (Fig. 33). This would accord well with the presence of tendonous muscles in all the surviving members of these groups.

Tendonous muscles may also perhaps have been evolved independently by most of the spiriferides, though no survivors exist to prove the point.

This interpretation of muscle platforms urgently needs to be tested by a thorough and comparative study of the muscle positions of

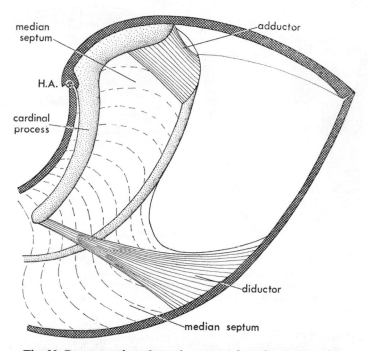

Fig. 30. Reconstruction of muscle system of an aberrant Permian strophomenide (*Scacchinella*; Strophalosiacea): note that the adductors are attached to an exceptionally large median septum and the diductors to an exceptionally long forked cardinal process.[87]

genera with and without platforms, and their relation to shell-form, giving particularly close attention to the anomalous platform-bearing genera in groups that generally lack platform structures.

Only in one articulate brachiopod is there any evidence that the normal system of leverage was modified significantly. Like many other exceptional features this occurred in a Permian strophomenide of 'coralloid' form (*Gemmellaroia*).[87] The lid-like dorsal valve has a massive cylindrical 'plug' fitting into a corresponding socket, and could only have moved vertically up and down (like the 'lid' of a rudist bivalve). This movement seems to have been powered by adductors and diductors with nearly parallel—but antagonistic—lines of action (Fig. 31).

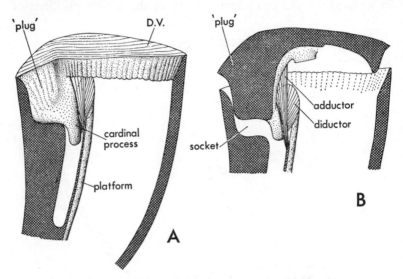

Fig. 31. Reconstruction of muscle system of a highly aberrant Permian strophomenide (*Gemmellaroia*; Strophalosiacea), A, lateral view of complete dorsal valve, with ventral valve cut along median plane; B, both valves cut medially, dorsal valve raised.[87]

Leverage in inarticulates

Most living inarticulates have no hinge of any kind and, by definition, no internal articulation. But when the shell opens the posterior valve edges may remain almost in contact while the anterior edges gape more widely; thus there may be a rotatory movement not unlike that of articulates.[11] Two main pairs of muscles extend more or less vertically through the coelom, and appear to act as *adductors* (Fig. 32). The anterior adductors are differentiated into two portions, possibly 'quick' and 'catch' in function, and may be homologous to the adductors of articulates. It is not entirely clear how the shell is opened. There is no obvious 'spring' between the valves, comparable to the ligament or resilium of a bivalve mollusc. But since the adductor muscles are massive, and the valves only gape apart by a few degrees, it is possible that the elastic recovery of the muscles

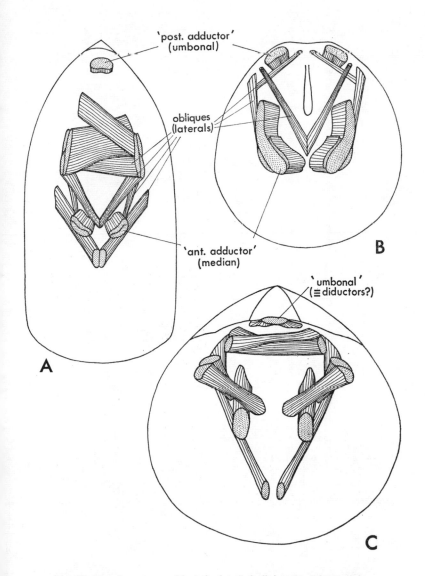

Fig. 32. Muscle systems of inarticulates: A, living *Lingula*; B, living *Discinisca*; C, reconstruction of Cambrian *Obolus*: dorsal views, dorsal valve removed. Muscle names and homologies as in Bulman:[18] names applied to *Lingula* in brackets. Note the 'umbonal' muscle reconstructed in *Obolus*, in position suggesting homology with the diductors of articulates.

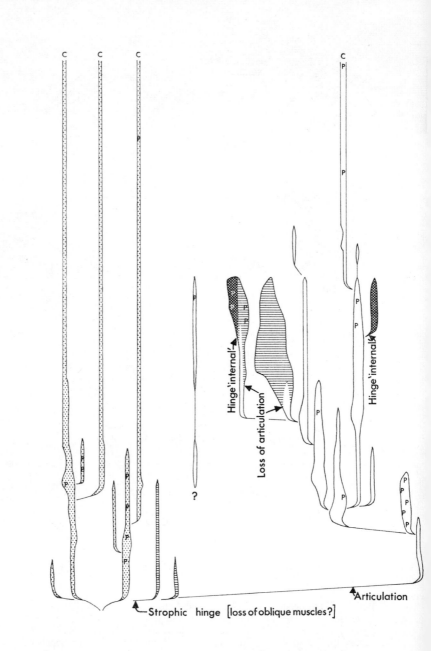

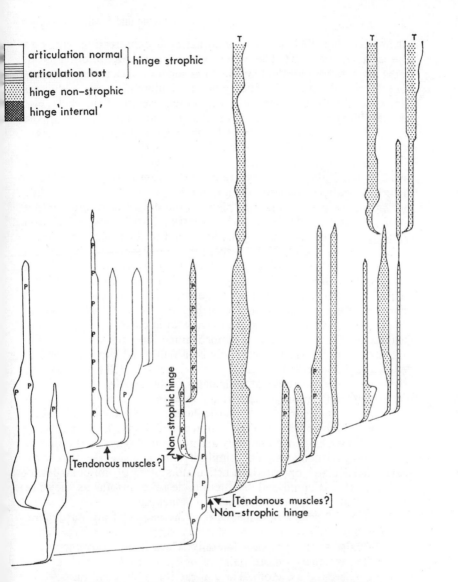

Fig. 33. Chart to show possible evolution of hinge and muscle systems in the Brachiopoda. Compare with Fig. 99; see also caption to Fig. 15. P, genera with muscle platforms (the latter is distributed evenly in those superfamilies in which platforms are general); C, extant groups with columnar muscles; T, extant groups with tendonous muscles.[86]

themselves is sufficiently strong to act as a compression spring to force the valves apart. The way in which the valves are observed to open by rotation does not support the suggestion that the coelom is used as a hydrostatic chamber. The inarticulate posterior adductors are generally situated very near the posterior edge of the shell (Fig. 32A,B). They have been reconstructed in the same position in the Cambrian lingulide *Obolus*, which suggests that they may be homologous to the diductors of articulates (Fig. 32C).

A crude articulation actually seems to have been present in a few extinct inarticulates but they are quite different in structure and age from the early articulates. For example, in one Silurian lingulide (*Dinobolus*) a postero-median 'plate' in the dorsal valve fitted into a corresponding single socket in the ventral. Clearly this articulation must have evolved independently from that of the true articulates.

Many fossil acrotretides (especially acrotretaceans) had fairly straight posterior valve edges which may have functioned as a rudimentary hinge line, as they are known to do in the surviving acrotretide *Crania*. Growth on these straight valve edges produced corresponding interareas on the shell. Some other Cambrian shells (paterinides, kutorginides) seem to show even closer analogies to the articulates. They have a well defined hinge line and interareas; the interareas are interrupted medially by a delthyrial gap more or less completely protected by convex delthyrial covers. (The similarity of these structures to those of articulates is masked by the usual unnecessary duplication of terms: an interarea becomes a 'pseudo-interarea' if it occurs on an inarticulate, and the delthyrial covers become a 'homeodeltidium' and 'homeochilidium', though there is no convincing evidence that they are not homologous to the similar structures of articulates.) This similarity strongly suggests that some early inarticulates operated an articulate system of muscular leverage, and this is confirmed by the muscle scars insofar as they are known. Only the internal articulation is lacking.

The lack of articulation means that the valves of an inarticulate are free not only to open and close by rotation around the hinge axis, but also to shear or slide laterally. These lateral movements are controlled by one or more pairs of *oblique muscles*, which extend obliquely across the coelom. The contraction of different pairs, or of one member of a pair, is capable of sliding the valves in various directions. These movements are used mainly to ensure that the valves fit one another accurately when the shell closes (Fig. 32B). But in *Lingula* the oblique muscles are exceptionally well developed, and include *transmedian* muscles crossing the median plane obliquely

(Fig. 32A); their sliding movements are used by this highly aberrant brachiopod in the maintenance of its burrow. Conversely the oblique muscles are much reduced in *Crania* with its rudimentary hinge line. The development of a true hinge line, and even more an articulation of teeth and sockets, would have made oblique muscles redundant, and so it is not surprising that they are altogether absent in articulates.

All living inarticulates have columnar muscles; a few fossil forms (e.g. trimerellaceans, some acrotretaceans), which developed highly convex shells, have structures resembling the muscle platforms of articulates. (The acrotretacean structures here interpreted as muscle platforms have been regarded as supporting structures for the lophophore,[109] but do not seem closely analogous to the brachidia of articulates.)

Summarising, the emergence of the first articulates from inarticulate ancestors could have happened quite gradually, if many of the early inarticulates were already operating 'articulate' muscular leverage. The inarticulate but clearly strophic paterinides and kutorginides are appropriate possible links between the crudely strophic acrotretides and the truly articulate early orthides (billingsellaceans) (Fig. 33).

4

RELATION TO SUBSTRATE

Structure of the pedicle

Throughout the history of the phylum the majority of brachiopods have been permanently attached to the sea bottom, or at least have rested permanently on it—except during a short free-swimming larval period. They are, and were, attached either by a pedicle or by cementation. Even those that rested freely on the bottom were anchored at least at some stage of growth. Of the two methods of attachment, the pedicle is certainly the more primitive.

At least among living genera, the pedicles of inarticulates and articulates are not homologous, and differ radically in structure, function and embryonic origin. In the articulates the larva settles on to the substratum after a short free-swimming period, and becomes attached by one of three larval segments. It is this segment that develops into the pedicle (Fig. 91A). In the inarticulates this larval segment is missing, the free-swimming period is longer, and settlement is delayed until a much later stage of growth (Fig. 91B), by which time a pedicle has developed from an enlargement of the posterior edge of the ventral mantle. This difference is retained even in cemented genera. The cemented articulate *Lacazella* is initially attached by a pedicle segment, though this soon atrophies and is replaced by the cementation of the ventral valve. The cemented inarticulate *Crania* never possesses even the rudiment of a pedicle, but becomes attached directly by cementation of the larval ventral valve.[69]

It is not at all clear how this difference should be interpreted, and our speculations are not aided by any fossil evidence. It has been customary to regard the inarticulates as the more 'primitive' in every respect, but it is equally possible that in this particular feature it is the articulates which have preserved the original character of the phylum. It cannot be assumed that all fossil inarticulates possessed the type of pedicle characteristic of their few living survivors. The loss of an original larval pedicle segment, and the acquisition of another pedicle formed later in ontogeny from quite different material, could have been connected with the secondary development of a longer free-swimming period.

The difference in origin of the two types of pedicle is reflected in their structure and function. The larval pedicle segment of an articulate grows into an organ that is quite distinct from either the body or the mantle. It is a tough solid cylinder, projecting through the posterior body wall and rigidly fixed to the substrate. It consists of a core of connective tissue, histologically similar to cartilage, surrounded by a layer of epithelium and a thick chitinous cuticle.[50] Within the coelom it is connected to the valves by dorsal and ventral pairs of unstriated *adjustor muscles* (Figs. 28, 34). These are inserted

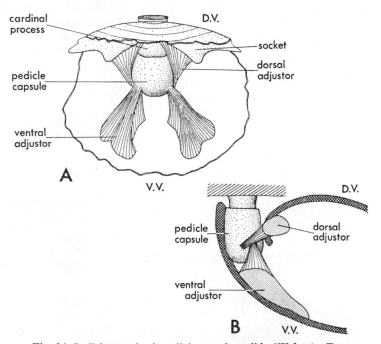

Fig. 34. Pedicle muscles in a living terebratulide (*Waltonia*; Terebratellacea): note that their oblique insertion into the pedicle capsule allows the shell to swivel in various directions. A, anterior view of a dissection of the muscles; B, lateral view of another dissection, valves cut along median plane.[86] See also Fig. 28.

into the inner end of the pedicle in such a way that the contraction of one pair, or of the left or right members of both pairs, can rotate the rest of the brachiopod dorsally or ventrally, or swivel it laterally. Most commonly they swivel the shell round through a wide arc just after it has snapped shut for the ejection of faeces; probably this facilitates the dispersal of the faeces by currents or wave-scour.[76] The pedicle itself is not muscular, and remains immobile throughout. The ventral adjustors are attached to the inner surface of the ventral valve, postero-lateral to the attachments of the diductors. The dorsal adjustors are often attached to the medial faces of the hinge sockets, to a hinge plate, or to some other part of the dorsal 'cardinalia' (Fig. 28). Adjustor muscles sometimes leave scars on fossil shells; their arrangement seems to have been fairly uniform throughout the history of the articulates. The enlarged socket walls (so-called 'brachiophores') of some orthides, which project forwards further than is necessary for articulation (Fig. 21), may have been used for attachment of dorsal adjustors. The ventral adjustors of some fossil brachiopods likewise seem to have been attached to the inner faces of the dental plates.

In living inarticulates, on the other hand, the posterior edge of the ventral mantle grows into a pedicle which retains its connection with the mantle and shares its structure. Like the mantle edge, the pedicle itself is highly muscular, and has no external pedicle muscles (Fig. 35). It is a hollow cylinder, with a central cavity derived from the coelom. There are strong muscles either in the cavity itself (*Discinisca*) or in its walls (*Lingula*), which are sheathed with epithelium and a chitinous cuticle. The central cavity has a sphincter valve at its

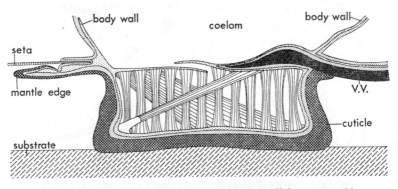

Fig. 35. Longitudinal section of pedicle of the living acrotretide *Discinisca*, to show complex musculature.[14]

connection with the coelom, and acts as a hydrostatic skeleton. The contraction of the various muscles within the pedicle serves to raise, lower or rotate the rest of the brachiopod.

Attachment of the pedicle

At the present day, the pedicles of both inarticulates and articulates are most commonly attached to some hard substrate, such as a pebble, a rock surface, or a piece of shell or coral. Generally the pedicle is short, so that the shell lies closely against the substrate. The chemical nature of the attachment is unknown, but it is extremely strong: it is easier to tear the pedicle away from the rest of the brachiopod than to detach it from the substrate. In some terebratulides (e.g. *Terebratulina*) the tip of the pedicle has a tendency to divide into short rootlets; these penetrate shell-fragments on the substrate in a manner that strongly suggests that they can dissolve the calcium carbonate.

Attachment to hard substrates was probably common among extinct brachiopods. Occasionally their shells are found with the foramen preserved closely against the surface of another shell or a piece of coral; evidently they were buried before the pedicle had decayed or currents had swept them out of their position of life. Such specimens are exceptional, however, and usually there is no direct evidence of the nature of the immediate substrate to which they were attached.

Very commonly fossil brachiopods are found in rocks which evidently accumulated as soft sediments, and there is no trace of any hard materials, except other shells, to which they could have been attached. In many cases the shells of dead individuals (or even those still alive) may have been sufficient to ensure the anchorage of each new generation. But this is not the only possible explanation, for some living brachiopods are known to be able to attach themselves in other ways. Many are able to attach the pedicle to 'soft' organic materials, such as the stems of algae, the tests of ascidians or the 'horny' tubes of polychaete worms. These materials would not normally be preserved in the fossil state, so that the shells would be buried in the sediment without any trace of their original attachment.[75] This is a likely explanation of the 'nests' of fossil brachiopods which are common in certain strata. These are dense patches of fossil shells in otherwise poorly fossiliferous sediments; biometrical analyses suggest that each 'nest' contains a true life-assemblage,[48] which implies that each may represent the population attached to a particular piece of unpreserved substratal material. Similar 'nests' have occasionally been found grouped around a pebble embedded in finer-grained sediment.

Very rarely, living brachiopods attach themselves directly into a soft sediment. The best example is a terebratulide (*Chlidonophora*) in which the pedicle rootlets already mentioned are developed to an exceptional length. When dredged from a deep ocean floor of *Globigerina*-ooze the rootlets are found to be penetrating the foraminiferal shells, and presumably the brachiopod is thus rooted into the soft ooze (Fig. 36). This example is important, because the foramen alone would give no hint that the pedicle is exceptionally long and rootlike. Therefore this method of attachment may have been much more common among extinct articulates than it is among living species. Among inarticulates, only *Lingula* is known to root itself into an unconsolidated sediment. It uses a sticky mucous secretion to attach the distal part of its long pedicle to the sand at the bottom of a deep vertical burrow; the contraction of the pedicle then allows it to withdraw into the burrow when danger threatens (Fig. 48A).

Pedicle foramen

The point of emergence of the pedicle may or may not be clearly reflected in the structure of the shell. If there is a foramen, its structure depends in part on whether the pedicle is of the articulate or inarticulate type.

The foramen for an inarticulate pedicle merely serves for the passage of the nerves, blood vessels and coelomic space; it may be much smaller than the pedicle it supplies. Thus in *Lingula* the pedicle emerges through a very small gap between the posterior edges of the valves (there is no true foramen), but outside the shell it swells to a much greater diameter. In *Discinisca* the pedicle initially emerges in the same way, but its point of emergence is gradually incorporated in a notch, and finally a slit, in the posterior side of the ventral valve; but this slit is much smaller than the pedicle, most of which is attached to the external surface of the ventral valve (Fig. 35). The foramen thus gives no direct evidence of the size of the pedicle.

This makes it difficult to reconstruct the pedicles of extinct inarticulates. Some of them (lingulides) have no true foramen. A narrow gap between the posterior valve edges might have been the point of

---

Fig. 36. A, Long branched pedicle of an abyssal terebratulide (*Chlidonophora*); B, enlargement showing rootlets penetrating shells of foraminifera.[25]

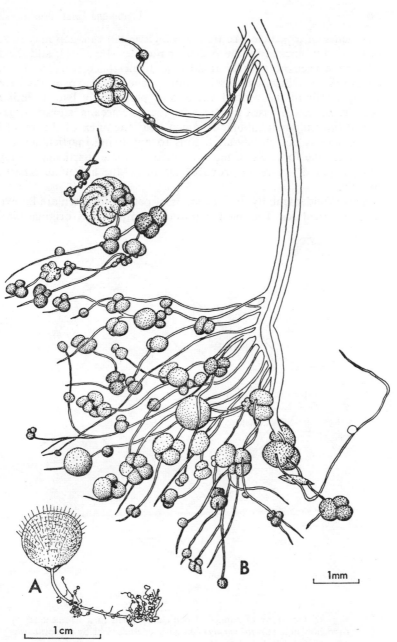

A

1cm

B

1mm

Fig. 62.

emergence of a pedicle like that of *Lingula*; but this gap may have been wholly blocked by the posterior muscles (Fig. 32c), and there may have been no pedicle at all (e.g. trimerellaceans). The fossil ancestors of *Discinisca* (discinaceans) generally have a posterior notch or slit in the ventral valve and presumably had a similar pedicle. But other fossil inarticulates (acrotretaceans, siphonotreta-ceans) have a small tubular foramen near the apex of the ventral valve. This is not closely analogous to any living inarticulate, but resembles the foramen of many articulates; these inarticulates may have acquired—or more probably retained—the articulate type of pedicle.

Three fundamentally distinct kinds of pedicle foramen are known among articulates. The most important and probably original kind

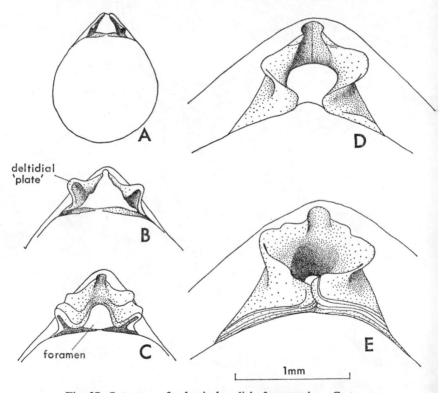

Fig. 37. Ontogeny of subapical pedicle foramen in a Cretaceous rhynchonellide (*Cretirhynchia*), showing encirclement by elaborate but unfused deltidial plates.[96]

may be termed *subapical* (='hypothyrid'), in that it pierces the ventral valve between the apex and the hinge, i.e. 'beneath' the apex of the isosceles-triangular delthyrium (Fig. 37). In the ontogeny of most articulates (other than strophomenides) the pedicle initially emerges between the posterior valve edges, and there is no special foramen. As the articulation develops and a delthyrial gap becomes necessary for the musculature, part of the rest of the gap is utilised for the emergence of the pedicle. Therefore, any delthyrial cover that develops to protect the remainder of the delthyrial gap is necessarily influenced by the presence of the pedicle. Generally the delthyrial cover takes the form of a pair of *deltidial plates* which encroach on

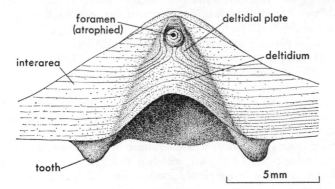

Fig. 38. Subapical pedicle foramen and deltidium of an Ordovician orthide (*Clitambonites*): note how the growth-lines show that paired deltidial plates gradually encircled the foramen during ontogeny before fusing into a deltidium; the foramen also became constricted, i.e. the pedicle atrophied.[86]

the delthyrium from either side, and so delimit a definite foramen (Fig. 37D). With further growth of the delthyrium they may join in the mid-line between the pedicle and the dorsal valve (Fig. 37E). If they then fuse, they form a *deltidium* (or 'symphytium') (Fig. 38). If the pedicle fails to grow in proportion to the whole shell, the foramen may be left ultimately as a small hole under the apex of a large convex deltidium. (The deltidium has traditionally been regarded as somehow closely related to the pedicle, but in fact its growing edge is spatially and anatomically separate from the pedicle and is related instead to the hinge: see p. 59).

Subapical foramina with deltidial plates or a fused deltidium are characteristic of the rhynchonellides both at the present day and throughout their history. They are also found in some orthides (clitambonitaceans and billingsellaceans). Most other orthides have no delthyrial cover, and therefore preserve no positive evidence of the emergence of a pedicle through the delthyrial gap. But a pedicle was very probably present, since deltidial plates developed sporadically in a few genera. Subapical foramina with plates or a deltidium are also found in some spiriferides (suessiaceans) and many atrypides. In some spiriferides with detached stegidial plates the plates were initially paired on the ventral side, and gradually enveloped a sub-apical foramen in a manner precisely analogous (though not homologous) to the growth of deltidial 'plates'.[33]

A subapical foramen suffers from one inherent functional disadvantage. At any growth stage the size of the pedicle is limited by the size of the delthyrial gap; and once it has been enveloped by deltidial or stegidial plates it cannot enlarge at all, and indeed must become relatively smaller as the shell continues to grow. Brachiopods with subapical foramina therefore cannot combine complete delthyrial protection with a strong pedicle attachment throughout ontogeny; and in fact many of them show signs of an absolute reduction—or even atrophy—of the pedicle during ontogeny (Fig. 38), the stability of the shell being maintained in other ways.

This inherent limitation was overcome in the development of the second major type of foramen, which may be termed *transapical*. In brachiopods with a transapical foramen the pedicle emerges initially in a subapical position, but migrates during ontogeny *through* the apex of the ventral valve, and simultaneously enlarges in size (Fig. 39). (Stages in this migration are generally termed 'mesothyrid', 'epithyrid', etc.; these terms may be useful for taxonomic description, but their multiplicity masks the important point that all of them involve the encroachment of the foramen through the ventral apex.) It is clear that this migration and enlargement must be due to *resorption* by the pedicle of the shell-material surrounding it: the development of a transapical foramen is therefore dependent on the ability of the pedicle epithelium to resorb calcite. It is probably significant that the living brachiopods (e.g. *Terebratulina*, *Chlidonophora*) with pedicle rootlets able to resorb calcareous material in the substrate are also among those with transapical foramina. Occasionally the pedicle becomes so large relative to the size of the shell that the dorsal apex is resorbed away as well as the ventral ('amphithyrid').[9] Clearly, with a pedicle able to resorb the surrounding shell there is no inherent limit to the size to which the pedicle can grow, and it can remain large enough to provide the shell with strong

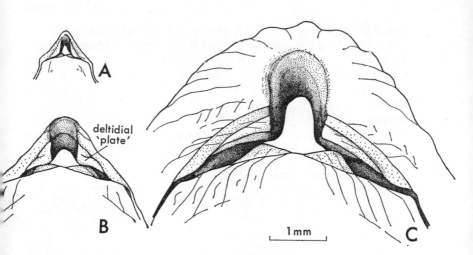

Fig. 39. Ontogeny of pedicle foramen in a Cretaceous terebratulide (*Gisilina*; Terebratulacea), showing enlargement and transapical resorption.[96]

anchorage throughout ontogeny. A transapical foramen also incidentally allows the shell to be even more closely pressed against the substratal surface, giving the pedicle itself still greater protection. With a transapical foramen, as with a subapical, part of the delthyrial gap may not be required for the foramen, and this is commonly filled with deltidial plates (Fig. 39) or a fused deltidium (Fig. 42).

A transapical foramen is characteristic of all terebratulides (except a few of the early genera) and of most members of two atrypide groups (retziaceans, athyridaceans), which probably include the ancestors of the terebratulides. No examples are known with certainty from any of the other orders.[34] It is therefore likely that the power of resorption in the pedicle, on which the development of a transapical foramen depends, was evolved only once, probably among the Silurian atrypides (Fig. 49). Considering the functional advantages of a transapical foramen, it is probably significant that the terebratulides, with such foramina, are by far the most abundant brachiopods surviving to the present day.

The third and last type of articulate foramen is the *supra-apical*, which is characteristic of all the strophomenides, except those branches of the order which later adopted cementation attachment. It is unknown in any other articulates, except possibly some primitive orthides (e.g. the ? billingsellacean *Matutella*). A foramen is supra-apical if at all growth-stages it pierces the ventral valve on the anterior side of the apex, rather than on the posterior or hinge-line side. Very young specimens of fossil strophomenides show that the foramen pierced the ventral valve extremely early in ontogeny, possibly even in the protegulum stage, and certainly before a hinge line was formed. Later in ontogeny, when a hinge line and interarea developed, the delthyrial gap necessary for the musculature could be protected by a 'pseudodeltidium', the growth of which was completely unaffected by the foramen. In most strophomenides the

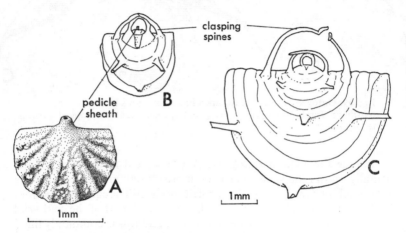

Fig. 40. Juvenile attachment of two strophomenides: A, an Ordovician davidsoniacean *(Fardenia)*, showing pedicle sheath around supra-apical foramen (from a photograph in [15]); B, C, growth-stages of a Carboniferous productacean *(Plicatifera)*, showing initial pedicle sheath superseded by tubular clasping spines.[16]

supra-apical foramen remained very small, though its edges were often extended outwards to form a tiny tubular *pedicle sheath* (Fig. 40A). It scarcely ever enlarged in proportion to the whole shell, and generally it must have become useless as a means of attachment after quite an early stage of growth (Fig. 40B, C). A very few strophomenides (e.g. *Leptaena*) have a slightly larger foramen; this suggests that the pedicle may have been able to enlarge during ontogeny by resorbing the shell-material around the edge of the foramen, and thus may have remained functional for a longer period.

In one small group of early inarticulates (siphonotretaceans) the foramen is somewhat similar, being supra-apical in position and showing clear evidence of resorption—in this case through phosphatic shell-material. This must be a quite separate development of resorbing power in the pedicle.

Cementation

A few groups of brachiopods have replaced pedicle attachment by a cementation of the ventral valve directly on to a hard substrate. The chemical nature of the cemented attachment, like that of the pedicle, is unknown; but it is very strong and survives fossilisation, and therefore presumably involves an intimate bond between the calcium carbonate of the shell and the material of the substrate. As already mentioned, in living brachiopods showing cementation this attachment is initiated at or very soon after spatfall. Fossil cemented brachiopods also show cementation even at the apex of the ventral valve, so that their cemented attachment must have been adopted equally early in ontogeny. It is therefore difficult to see how cementation could have developed unless paedomorphically.

The fossil record in fact suggests that cementation has only been evolved a few times. It first appeared among the inarticulates (craniaceans—but not some of the earliest members) in the Ordovician period. This group has continued without major change to the present day (*Crania*), and apart from a few doubtful cases it includes all known cemented inarticulates. Among the articulates cemented shells are confined to the strophomenides.[36] Cementation appeared for the first time somewhat later than among inarticulates, during the Silurian period, in a very few members of one strophomenide group (strophomenaceans). A different group (davidsoniaceans), which had existed since the Ordovician, seems to have developed cementation in the Devonian period: these brachiopods, with some aberrant derivatives (lyttoniaceans), were important throughout the later Palaeozoic era, continuing as a minor offshoot (thecideaceans) even to the present day (*Lacazella*). A third group of cemented strophomenides (strophalosiaceans) first appeared in the Devonian period, but seems unrelated to any other cemented group. It too was abundant throughout the later Palaeozoic era, also giving rise to aberrant offshoots (richthofeniids); but none of these survived the end of the Palaeozoic. In all, four separate developments of cementation can be identified in the phylum with some confidence, but there may have been a few others (Fig. 49).

Ecologically a cemented brachiopod would be at a disadvantage relative to one attached by a pedicle, in that it could only settle on hard substrates. But compensating for this, perhaps, would be its greater strength of attachment and consequent ability to colonise environments with strong wave or current action. It might also be better able to resist certain forms of predation.

Only in a few cemented brachiopods does the ventral valve edge remain in contact with the substrate throughout ontogeny. Generally, at an earlier or later growth stage, the valve edge rises off the sub-

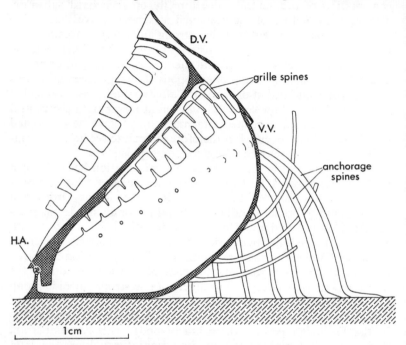

Fig. 41. Complex spines on a Permian strophomenide (*Chonosteges;* Strophalosiacea): note early attachment by cementation, later supplemented by tubular anchorage spines (based on photographs in [60]). Other spines may have been sensory.

strate, so that the attachment becomes confined to an area of variable size at the ventral umbo (Fig. 27). If the shell is broken off, this area is seen as a scar or *cicatrix*. In one of the groups of cemented strophomenides (strophalosiaceans) the attachment was supplemented by the cementation of tubular spines (Fig. 41). These were formed as outgrowths from the ventral valve edge; they grew in length until they encountered the substrate or an adjacent shell, and then grew along it, attaching to it by cementation. In some of these strophomenides the adult shell was attached by a dense thicket of these cemented root-spines, the original cementation of the shell itself being greatly reduced in importance. In some late Palaeozoic species these shells even grew in quite massive reef-like clusters.

Free-lying brachiopods

As already mentioned in passing, the original attachment by a pedicle or by cementation is very frequently reduced in importance during ontogeny. A pedicle may cease to grow in proportion to the whole brachiopod; or it may be unable to grow because it is confined within a foramen that it cannot enlarge by resorption. Becoming relatively smaller, it may no longer be able to support the shell. Then it may merely tether the shell in one position on the substrate, preventing it from being moved around by currents but unable to support its weight. Some living articulates (e.g. *Neothyris*) have pedicles that are reduced to this tethering function in the adult stage. Similar shells, with proportionately very small adult foramina, are extremely common among fossil brachiopods (Figs. 38, 42). In many of them the pedicle must ultimately have atrophied altogether. This is not known with certainty in any living species. But in many fossil shells it can be seen that the foramen was reduced in diameter and finally plugged by the deposition of shell-material; alternatively the umbo of the valves became so incurved that no passage for the emergence of a pedicle would have remained (Fig. 58). Thus it is certain that many fossil brachiopods in their later stages of growth must have lain freely on the sea-floor. Occasionally such shells may be preserved in large numbers in what was evidently their natural position of life.[115]

A similar ontogenetic sequence can be traced in many fossil brachiopods which were attached initially by cementation. The ventral valve edge grew off the substrate at some stage, and thereafter the area of cementation could not increase. Becoming relatively smaller, it must often have become inadequate to support the gradually enlarging shell. In some species it remains in the adult stage only as a tiny cicatrix at the ventral apex of a large shell. Possibly the shell sometimes broke free under its own weight; but if it was initially attached only to some small shell-fragment lying in a soft substrate, it would merely have heeled over as it outgrew the attachment.

Most brachiopods which became free-lying developed other means of stabilising their position on the sea-floor. Without such features, the shell would be liable to be tilted or inverted by currents or the

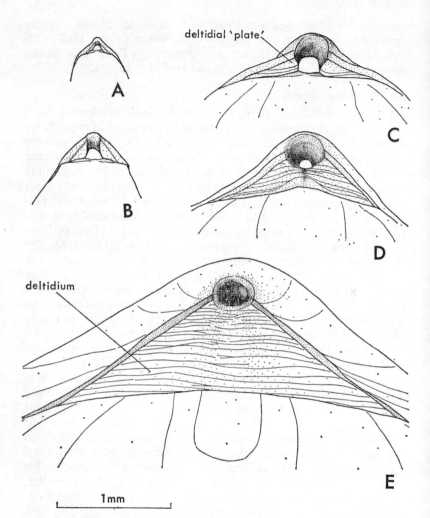

deltidial 'plate'

A

B

C

D

deltidium

E

1mm

Fig. 42. Ontogeny of pedicle foramen in a Cretaceous terebratulide (*Chatwinothyris,* Terebratulacea), showing growth of deltidial plates (A–C) and their fusion into a deltidium (D, E), enlargement and transapical resorption of foramen (A–C) and its later contraction (D, E).[96] The adult (not shown) had a relatively minute foramen, and was probably free-lying.

movements of other animals; its valve edges might be buried in a soft sediment and its feeding and respiration thereby inhibited.

The simplest of these stabilising features is the mere weight of the shell. As the shell grew, it would become less liable to disturbance by chance currents. Generally, however, extra weight was added progressively on the posterior side of the shell-cavity by a great thickening of the secondary (or prismatic) layer. This differential weighting held the shell in a stable position, with the posterior side downwards and the apertures away from the substrate. Any slight displacement would automatically be righted, just as a weighted toy figure automatically returns to an upright position. In fact this posterior weighting is common even in shells with a functional pedicle; and as the pedicle becomes reduced to a mere 'tether', so the importance and extent of the weighting increases. Posterior weighting is very common among extinct articulates, and has without doubt been evolved on many separate occasions.

Stability seems to have been achieved in a different way by some spiriferides, in which the ventral interarea developed into a broad flat base for the shell (e.g. *Cyrtia*, *Cyrtina*). In at least one genus (*Syringospira*) the area of this base was increased still further by the lateral extension of the interarea into a pair of thin sheets of shell-material (Fig. 43).[28] Moreover, much of the extra volume of shell-cavity resulting from the lengthening of the ventral valve was sealed off internally by irregular shelly partitions, and was evidently not required for an enlarged mantle cavity or any other internal function.

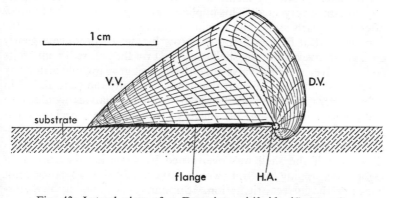

Fig. 43. Lateral view of a Devonian spiriferide (*Syringospira*; Spiriferacea) with highly extended plane interarea on ventral valve, extended laterally into sheet-like flanges (here seen edge-on), and possibly stabilising the shell on a soft substrate.[28]

This makes it probable that the stability of the shell was the primary function of all these broad high plane interareas. Like the posterior weighting, these flat bases originally supplemented a functional pedicle, but were often developed further when the pedicle was reduced or lost. Some other spiriferides (e.g. *Mucrospirifer*), although not having a flat interarea, extended the hinge line laterally into a pair of slender *alae*, which may have had a ski-like function to stabilise the shell on a substrate of soft sediment (Fig. 44).[90] (These alae have a superficial appearance of tubular 'siphons', but they are closed distally and are too close to the hinge axis to have functioned as siphons.)

Most of the strophomenides adopted a quite different means of keeping the valve edges away from the substrate after the atrophy of

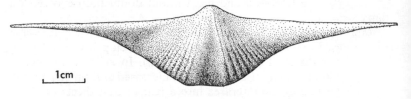

1cm

Fig. 44. Shell of a Devonian spiriferide (*Eleutherokomma*) with greatly extended ('mucronate') hinge line, which possibly served to stabilise the shell on a soft substrate (based on photograph in [37]).

the pedicle. As already mentioned, the supra-apical foramen generally remained very small. This implies that the pedicle ceased to grow after a very early stage, and would soon have become inadequate for anchorage. But by modifying the growth rates to produce a gently concavo-convex form, the shell could rest on the soft sediment on its convex valve, while the valve edges were kept growing upwards away from the substrate. The convex valve is generally the ventral, but in a few genera it is the dorsal instead: both alternatives would have had the same effect (Fig. 24). Shells of this kind are especially common among the earlier strophomenides (plectambonitaceans, strophomenaceans, chonetaceans). Experiments with working models show that if the shell was overturned by some bottom current, a vigorous snapping reaction would have enabled it to somersault back into the correct orientation.[86] Similarly, if sedimentation threatened to clog the valve edges, a snapping reaction would have caused the whole shell to rise off the substrate and move posteriorly (Fig. 45).

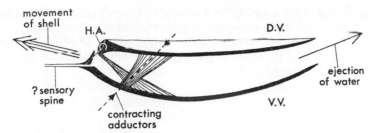

Fig. 45. Reconstruction of possible swimming mechanism of a gently concavo-convex strophomenide (*Chonetes*): a contraction of 'quick' adductors would eject water anteriorly and therefore thrust the shell posteriorly but upwards.[86]

Swimming brachiopods

Such use of a snapping reaction to avoid sedimentation is closely analogous to the escape reaction used by many free-lying pectinid molluscs at the present day; and some of these gently concavo-convex brachiopods may have used it likewise to escape predators as well as sedimentation. But they could not have swum forwards like the pectinids, because there is no evidence that the mantle edges of any brachiopods were ever modified into the vela which are essential for that swimming mechanism. But by snapping the shell repeatedly it is possible that they could have swum in a posterior direction. It is unlikely that any brachiopods were truly nektonic, but some may have swum short distances near the bottom. Some of the strophomenides with small, thin, gently concavo-convex shells (especially chonetaceans) would have been well adapted to such a mode of life (Fig. 45).

Quasi-infaunal brachiopods

Many of these strophomenides again altered their growth rates later in ontogeny, so that the shell became much more strongly concavo-convex. This is often combined with the development of great differential thickening of the convex (generally ventral) valve. This would have stabilised the shell against overturning, and if the sediment was soft the convex valve may have partially sunk into it. If sediment settled into the concave valve above, the shell might have been concealed except for the crescentic valve edges projecting above the surface of the sediment (Fig. 47).[26, 46] This would be the nearest to truly infaunal habit that any articulates have ever achieved.

Strongly concavo-convex shells, which might have had this quasi-infaunal habit, are common in most of the strophomenide groups.

The change from a gently to a strongly concavo-convex form was often quite sudden in ontogeny, being reflected in a sharp change in the concavity of one valve and, less commonly, in the convexity of the other also (Fig. 8E). On the present interpretation this would

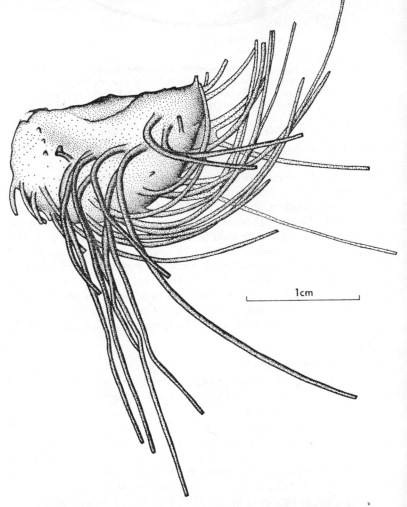

Fig. 46. Lateral view of a Permian strophomenide with tubular spines on ventral valve (*Marginifera*; Productacea): these spines probably rooted the shell in a soft substrate (from a photograph in [47]).

represent a fairly rapid switch from a free-lying, mobile, and perhaps swimming mode of life, to a static and more-or-less infaunal position. This would not have been the only major functional change during ontogeny, since all these shells show signs of early attachment by a pedicle or by cementation.

Some of the later strophomenides (productaceans, strophalosiaceans) seem to have developed this mode of life still further, for the shells were stabilised by a variety of devices formed from projecting tubular spines.[60]

Early in ontogeny, spines grew from the ventral valve on either side of the tiny pedicle and clasped the shell to some piece of substratal material (Fig. 40B, C). This effectively prolonged the attached stage, but except in rare instances[44] even this attachment generally became inadequate as the shell continued to grow in size. Later spines were therefore adapted to stabilise the shell on or in the substrate. In some genera, clusters of spines seem to have penetrated like roots into the soft sediment (Fig. 46). Similar spines on the concave dorsal valve could have held in position a mass of camouflaging sediment. In other genera a few long stout spines, or a thicket of finer spines, project in particular positions, spreading out horizontally around the shell: possibly these spines penetrated a very soft sediment a little below its surface and prevented the shell from sinking too deeply (Fig. 47). Brachiopods with these various stabilising

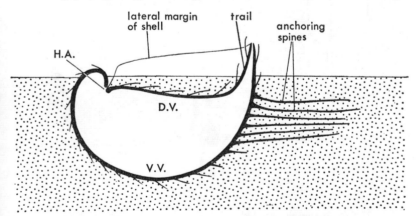

Fig. 47. Reconstruction of living position of a Permian spiny strophomenide (*Waagenoconcha*; Productacea), shown in longitudinal section. Note the dense thicket of large spines (also extending laterally) near the inferred surface of the substrate, and the 'quasi-infaunal' position of the whole shell (modified from [46]).

devices were very abundant throughout the later Palaeozoic, though none survived the end of the era: their development was of course dependent on the ability of the mantle to secrete tubular spines.

Burrowing brachiopods

Without the development of erectile mantle edges or, still better, mantle edges that could be fused and produced into siphons, brachiopods could not develop a truly infaunal habit comparable to that of

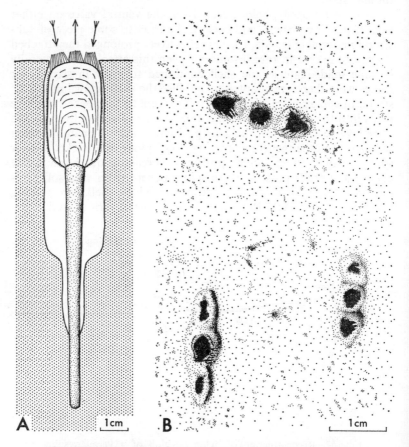

Fig. 48. Burrows of the living *Lingula*, with shells in feeding position: note three siphon-like apertures. When disturbed, shell is withdrawn by contraction of pedicle. A, section of burrow;[21,86] B, three burrows as seen at surface of muddy-sand substrate.[86]

many bivalve molluscs. No brachiopods, probably, have ever developed the property of mantle fusion which would have enabled them to form siphons. But one group of living inarticulates (lingulaceans) has erectile mantle edges which, with the aid of close-set chitinous *setae* (see p. 100), are used to form structures remarkably similar to the fused mantle edges and siphons of a bivalve. Without doubt it is this that has enabled *Lingula* and related genera to exploit a mode of life which is altogether exceptional among the brachiopods. As mentioned earlier, lingulids form a deep vertical burrow, attaching the long muscular pedicle to the sand at the bottom of the burrow (Fig. 48A). The upper part of the burrow, down which the shell can be withdrawn if the animal is disturbed, is maintained by rhythmic movements of the setae on the long parallel sides of the shell, and by the copious secretion of mucus from the mantle edges there. When the valves are open the setae are also erected across the gape at the sides of the shell. Anteriorly, at the upper end of the shell, the setae are moulded into the form of three short siphons. Only these pseudo-siphons project above the substratal surface (Fig. 48B).

This exceptional burrowing habit can be traced as far back as the Ordovician period, for lingulid shells are sometimes preserved upright in the strata. It is a mode of life that has apparently remained constant for an astonishing period of time (about 450 million years). It cannot be assumed, however, that all lingulides were burrowers. Many, especially the earlier ones, lack the characteristic parallel-sided shell-form of those known to be burrowers. They may have had a more normal epifaunal mode of life, though perhaps using their muscular pedicle to anchor the shell in a soft substrate.

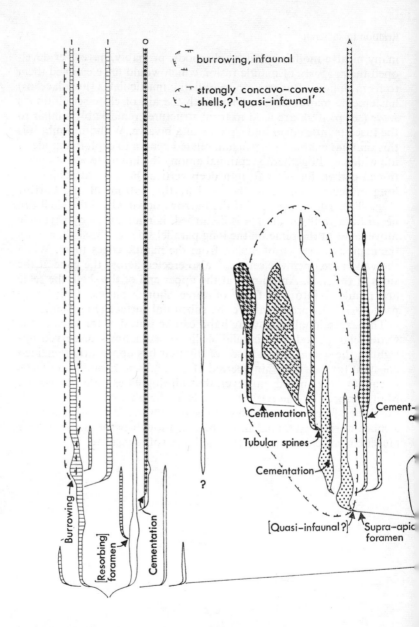

burrowing, infaunal

strongly concavo-convex
shells,? 'quasi-infaunal'

Cementation

Tubular spines

Cementation

Cement-
a

[Quasi-infaunal?]

Supra-apic
foramen

Burrowing

[Resorbing]
foramen

Cementation

?

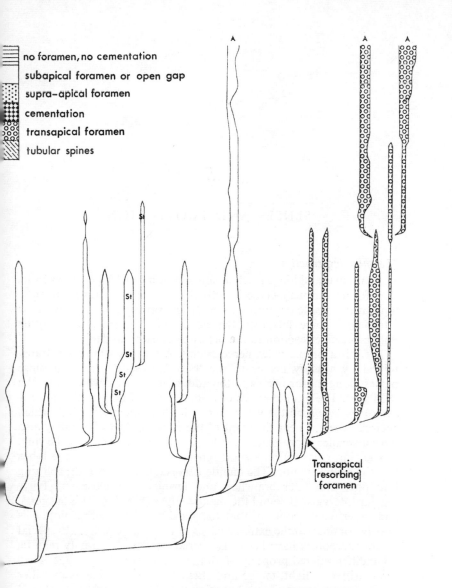

no foramen, no cementation
subapical foramen or open gap
supra-apical foramen
cementation
transapical foramen
tubular spines

St
St
St
St
St

A

A

A

Transapical
[resorbing]
foramen

Fig. 49. Chart to show possible evolution of pedicle structures and modes of attachment to the substrate. Compare with Fig. 99; see also caption to Fig. 15. St, stegidia; A, larval pedicle of 'articulate' type; I, larval pedicle of 'inarticulate' type; O, no larval pedicle.[33,34,86]

5

SENSES AND PROTECTION

The snapping reaction

The muscular and hinge mechanisms enable a brachiopod to open its shell for its various metabolic activities or to close it for protection. Since it is a sessile organism, more or less permanently resting on or attached to some substrate, the closing of the shell is virtually its only protective reaction to the presence of external danger.

There is a simple reflex nerve circuit. A sensory stimulus is transmitted by sensory nerves to the 'brain', from which an impulse passes along motor nerves to the adductor muscles, which snap the shell shut. The brain consists of a small mass of nerve cells, the *subenteric ganglion*, on the ventral side of a 'circumenteric ring', which encircles the gut near the front of the coelom. Articulates may also have a smaller supra-enteric ganglion on the dorsal side of the ring.

The ring gives off important sensory nerves to the mantle lobes and the pedicle region. The mantle nerves radiate outwards, splitting into finer and finer branches, and terminate at close and fairly regular intervals all round the mantle edges (Fig. 50). This suggests, and experiments confirm, that the sensitivity of the brachiopod is largely confined to the extreme edges of the mantle lobes. No special sensory receptors have been discovered, and the sensitivity may be an undifferentiated property of all the cells at the mantle edge. They are sensitive to light, tactile and chemical stimuli. Like many other sessile invertebrates, most brachiopods show a highly developed 'shadow reflex', snapping the shell shut whenever there is a sudden fall in the intensity of light. There are no 'eyes' or pigment spots at the mantle edge, and the light sense is presumably dermal. There is no reason to suppose that the mantle caecae are involved in any way: they are not supplied with nerves, and in any case the shadow reflex is as well developed in some impunctate as in punctate species. The mantle edge is extremely sensitive to tactile stimuli: the shell snaps shut if either edge is lightly touched by some small swimming animal or by a drifting sand grain.[77] The nature of the chemical stimuli is not well understood; but some brachiopods are known to react to the presence of hyposaline or highly turbid water, and it is possible that potential predators may be detected at a distance.

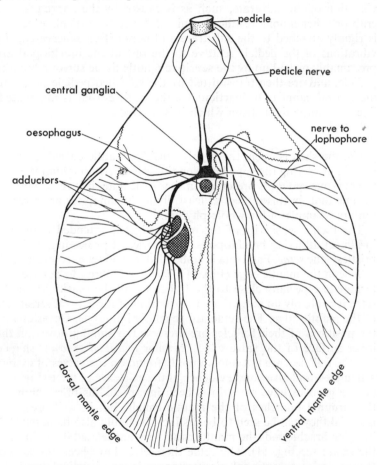

Fig. 50. Nervous system of a living terebratulide (*Magellania*; Terebratellacea): dorsal view of decalcified specimen, dorsal mantle removed on one side; note sensory nerves terminating around mantle edges, and motor nerves to adductor muscles.[49]

Whatever the nature of the stimulus, an impulse must pass rapidly along the mantle nerves to the central ganglia, and back to the 'quick' posterior adductors; for they contract immediately and snap the shell shut. The slower contraction of the unstriated anterior adductors is then able to hold the valves firmly together, if necessary for periods of an hour or more.[74] The reflex in an inarticulate is probably analogous.

Most articulate brachiopods react in the same way to any slight vibration of the substrate, such as is caused by the approach of a crab or other mobile benthonic animal. Since the articulate pedicle is rigidly attached to the substrate and is not itself innervated, the vibrations of the pedicle relative to the rest of the brachiopod are presumably picked up by the sensitive mantle tissue surrounding the pedicle, and are then transmitted by the so-called pedicle nerves to the central ganglia. In inarticulates the pedicle itself is innervated, like the mantle tissue from which it is derived.

The setae

While the shell is open, the highly sensitive mantle edges stand 'on guard', bordering the gape between the valves. Although no trace of the nervous system is preserved in the fossil state, it is reasonable to assume that, in this respect at least, living brachiopods are a representative sample of the whole phylum.

The tactile sensitivity of the mantle edges is extended outwards, in almost all living brachiopods, by a series of projecting chitinous bristles—the *setae*. Each seta is embedded in, and secreted by, a small follicle in the mantle edge (Fig. 51). The nerve fibres do not seem to terminate especially close to the follicles, and the sensitivity of the setae is probably no more than an extension of the general sensitivity of the mantle edge. The setae are generally immobile. Occasionally, the positions of their follicles are preserved as small grooves on the inner surface of the valve edges. Similar *marginal grooves*, shaped like the follicles, are not uncommonly found on the valves of extinct articulates, and give good evidence of the existence of setae (Fig. 52). (These grooves have sometimes been interpreted as impressions of the terminal branches of the mantle canals, but in well preserved material they can be seen to be distinct; it should also be noted that in living brachiopods the mantle canals are not related in position to the setae: see Fig. 51). The setae themselves have been found preserved under exceptional circumstances in one inarticulate of the middle Cambrian period.

During the growth of a brachiopod its mantle edges become longer, and the setae would therefore tend to become more widely spaced.

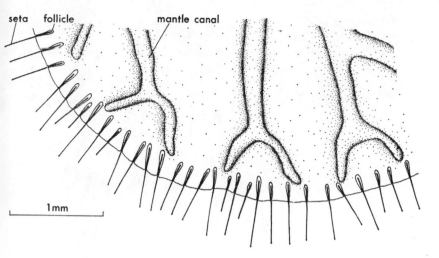

Fig. 51. Mantle edge of a living rhynchonellide (*Notosaria*), showing setae in follicles, and blind-ended mantle canals (setae not preserved to full length).[86]

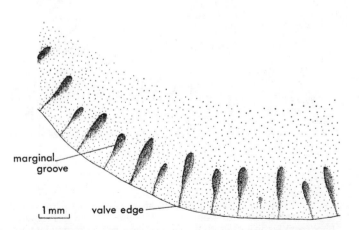

Fig. 52. Marginal grooves in valve edge of an Ordovician orthide (*Estlandia*: Clitambonitacea): note that they are pear-shaped like setal follicles.[86]

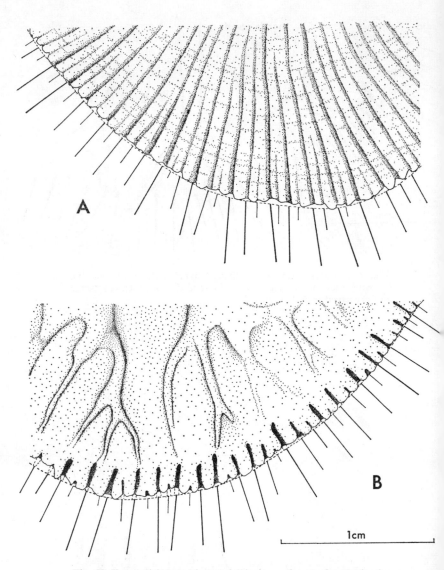

Fig. 53. External (A) and internal (B) views of part of ventral valve of an Ordovician strophomenide (*Palaeostrophomena*; Plectambonitacea), showing internal marginal grooves corresponding to external radial costellae. Setae are reconstructed (lengths arbitrary but proportional to lengths of grooves): note intercalation of new setae, like new costellae, mid-way between older ones.[86]

In fact, however, the spacing of the setae around the mantle edges is kept fairly constant by the continual formation of new setae to fill the gaps (Fig. 57). In living brachiopods there are two distinct ways in which new setae may be formed. A new seta either makes its appearance within the follicle of an older seta, and then branches off laterally forming its own follicle; or else it appears with its own follicle mid-way between two older setae. Occasionally (e.g. *Terebratulina*) the setae correspond precisely in position to the interlocking crenulations of the valve edges, and the pattern of fine radial ridges or *costellae* on the valve surfaces, formed by the crenulations during ontogeny, then corresponds exactly to the history of the setae. If new setae are formed by branching from older setae, new costellae will appear on the valve surfaces by branching from older costellae. The constant spacing of the setae around the mantle edges throughout ontogeny is precisely reflected in the constant spacing of the costellae on the valve surfaces.

Similar patterns of radial costellae are very common on the shells of extinct articulates. In many of them it can be confirmed that the crenulations of the commissure correspond exactly to the positions of marginal grooves within the valve edges (Fig. 53). The radial costellae show that, as in living species, the spacing of the costellae was kept constant during ontogeny by the formation of new setae, either by branching (Fig. 54) or by intercalation. The branching

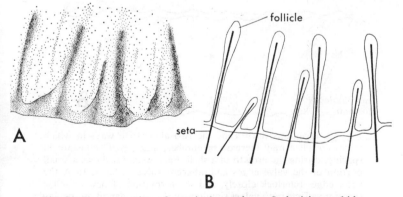

A

B

Fig. 54. Reconstruction of marginal setae in an Ordovician orthide (*Onniella*; Enteletacea). A, small portion of inner surface near valve edge, showing marginal grooves underlying external costellae, with interlocking ridges between; B, reconstruction of setae in grooves (not shown to probable full length), newer setae having branched from follicles of earlier setae (costellae are of 'branching' type).[86]

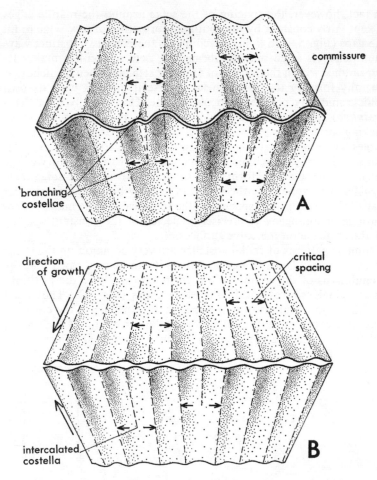

Fig. 55. Diagrams illustrating the two alternative ways in which radial costellae may increase in number, and remain constant in spacing, during the growth of a shell. Each diagram shows a small portion of the valve edges and adjacent valve surfaces. In A the valve edges interlock closely, and the formation of new costellae therefore affects both valves simultaneously, by apparent 'branching' from the flanks of an older costella. In B the costellae are entirely independent on the two valves, and new costellae are formed by intercalation between older ones. In both cases new costellae are formed whenever the spacing between earlier ones reaches a critical value, shown by the arrows.[85]

pattern was especially common among the orthides during the Palaeozoic era; but it recurred from time to time in other orders. The pattern of intercalation occurred in some of the earlier orthides but was most characteristic of the strophomenides; it too is found occasionally in other orders. These patterns can be regarded as functionally equivalent ways of achieving the same end (Fig. 55).[85]

Setae were undoubtedly more common than the occurrence of either costellae or grooves would imply, for many living brachiopods have setae that leave no trace on the valves either internally or

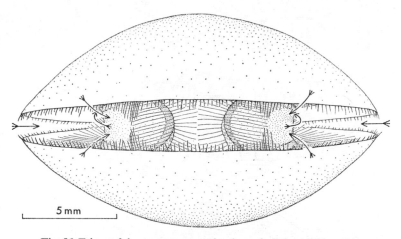

5 mm

Fig. 56. Fringe of short setae on mantle edges of a living terebratulide (*Neothyris*; Terebratellacea); anterior view, lophophore visible within. Note that the setae are longer at the junctions between the apertures, giving some protection to the apertural filaments.[86]

externally. But, conversely, not all types of crenulation and radial costellae correspond in position to setae, either in living species (Fig. 57c) or, probably, in fossils. A few living articulates (megathirids, thecideaceans) and inarticulates (craniaceans) have no setae at all.

In some living brachiopods the setae are closely spaced and short: they merely form a sensitive 'fringe' along the edges of the gape (Fig. 56). But in other genera (e.g. *Notosaria*, *Terebratulina*, *Megerlia*) they are much longer and more widely spaced; when the shell is open they straddle the entire gape with a kind of sensitive grille (Figs. 57, 80c). (Setae are brittle and are generally broken off in dredged specimens: such specimens may give a highly misleading impression of the importance of setae.) The extinct articulates in which the traces of setae are most clearly preserved were probably some that had setal grilles of this kind. Shells with radial costellae, which might have belonged to this category, are found among the first known articulates (e.g. *Nisusia*), in the early Cambrian period. Some early strophomenides (plectambonitaceans) have radial costae which probably reflect the positions of setae but are very widely

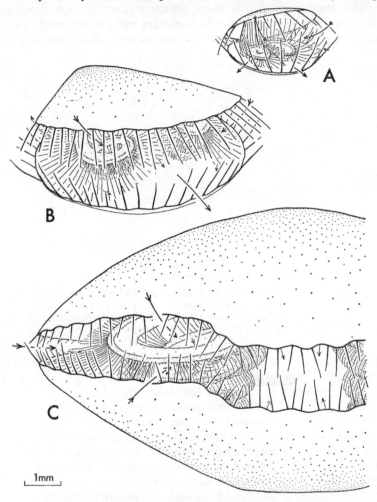

Fig. 57. Ontogeny of setal grille in a living rhynchonellide (*Notosaria*), showing constancy of spacing of setae. The lophophore is visible within the mantle cavity.[86]

spaced. Here the setae may have been few but long, projecting outwards like antennae.

Sensory spines

As means of protection setae suffer from the inherent disadvantage that they confer only tactile sensitivity. Various other devices seem to have been adopted by many fossil brachiopods, to utilise the wider sensitivity of the mantle edges themselves. Assuming that the valve edges were accompanied by (and secreted by) the sensitive mantle edges, many modifications of the valve edges can best be interpreted as devices that served to improve the efficacy of the snapping mechanism as a protective reaction.

For example, if parts of the sensitive mantle edges could be extended outwards beyond the rest of the shell, they could give 'early

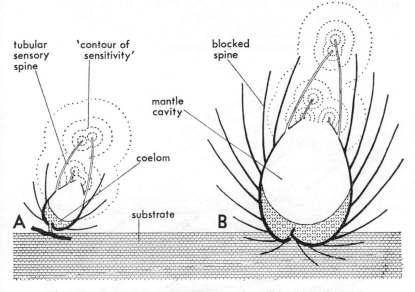

Fig. 58. Spines on a Jurassic rhynchonellide (*Acanthothiris*), illustrating their probable protective function: A, juvenile; B, adult. At each growth stage, only the most recently formed spines remained tubular, earlier spines being blocked. The tubular spines would have contained mantle tissue, and their tips could therefore have functioned as points of sensitivity (arbitrary 'contours of sensitivity' are marked around them). Note that in B some of the earliest spines may have functioned secondarily for stabilising the shell in a soft substrate, the original pedicle having atrophied.[83]

warning' of the approach of potentially harmful agents. Tubular spines are some of the most striking means by which this could have been achieved. Each was formed during ontogeny by differential growth at the valve edge, whereby a small ring of mantle-edge tissue was 'budded off' from the rest and thereafter grew independently, secreting a tubular spine. This piece of mantle tissue would presumably have retained its sensitivity, so that the spine could have functioned as a sensitive 'antenna' as long as it retained communication with the rest of the mantle.[83]

In many spiny brachiopods the spines were formed at regular intervals all round the valve edges, and would have carried the sensitive mantle edges outwards from the gape. But during the growth of the shell each spine would have been left further and further from the valve edge, and would have become less effective for protecting the gape. New sets of spines were therefore formed continually during ontogeny, the older and 'useless' ones being abandoned and sealed off at the base by shell material (Fig. 58). At any given stage of growth, only those nearest the valve edges would have been functional. Sensory spines are generally slender, and often project from the

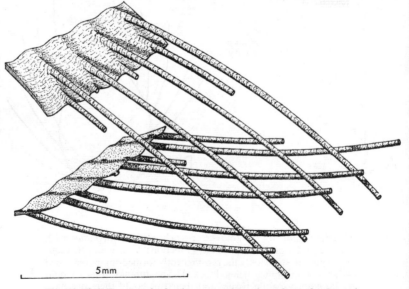

5mm

Fig. 59. Tubular marginal spines near the valve edges of a Jurassic rhynchonellide (*Acanthothiris*), showing how they could have straddled the gape as a protective grille, new spines being formed in each row as the shell grew.[83]

valve surfaces at a low angle, so that the most recently formed would have been well placed to provide 'early warning' protection to the gape between the valve edges (Fig. 59). Among the later strophomenides, these slender 'prostrate' spines often co-exist with the stouter spines ('rhizoid' or 'halteroid') which, as already mentioned, were well adapted to anchor the ventral valve on or in the substrate (Figs. 41, 47).

In the earliest shells in which tubular spines occur (chonetaceans), they project backwards in a single row from the broad posterior edge of the shell. In this position they would have carried portions of

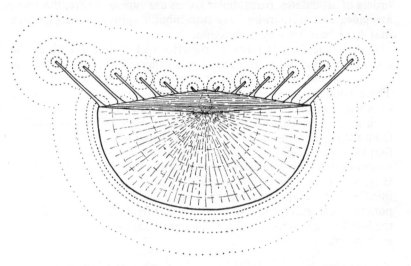

Fig. 60. Possible sensory function of tubular spines in a chonetacean (strophomenide), assuming retention of sensitive mantle edge at tips of spines: showing hypothetical 'contours of sensitivity' around entire shell. Dorsal view.[86]

the sensitive mantle edges outwards from the shell in this otherwise unprotected direction (Fig. 60). These brachiopods are some of the lightly constructed, gently concavo-convex forms which were probably free-lying and may even have been (in a limited sense) swimmers: if so, their row of sensitive spines would have projected in a functionally anterior direction during escape movements or swimming (Fig. 45).

Once these brachiopods adopted the quasi-infaunal mode of life, some of the spines would have acquired the secondary function of stabilisation and anchorage in addition to the original sensory function, and they could then have diversified in form.

The formation of these tubular spines depends on the ability of the mantle-edge tissue to re-fuse with itself during the 'budding off' process. Curiously, this does not seem to have been evolved many times (Fig. 63). Tubular spines were first developed in the Silurian period, in the concavo-convex brachiopods (chonetaceans) mentioned above. This group probably gave rise to other spiny strophomenide groups (productaceans, strophalosiaceans), but none of these abundant spiny brachiopods survived the end of the Palaeozoic era. In other orders of articulates, true tubular spines are curiously rare. But some atrypides and spiriferides have non-tubular spines, some of which may have been sensory in function.

A few brachiopods formed projections of a different kind: the commissure (i.e. both valve edges) was deflected outwards at certain points, forming narrow spikes projecting from the rest of the shell. These would have carried certain points on the mantle edge outwards in advance of the rest.[82] Three times in the history of the articulates such shells evolved with two symmetrical pairs of spikes radiating from the edge of the shell. These highly distinctive shells are so alike that they were originally placed in the same species; but there is now no doubt that their similarity is a spectacular case of homoeomorphy (Fig. 98). One genus (*Tetractinella*) is an impunctate atrypide of middle Triassic age; the others (*Cheirothyris, Cheirothyropsis*) are punctate terebratulides, and lived about fifty million years later in the late Jurassic period, and yet apparently are not closely related to each other.

Zig-zag slits

Another group of protective devices, while not providing 'early warning', would have improved the quality of protection at the gape itself. A wide gape may be necessary, or at least advantageous, for metabolic activities; but it increases the distance between the sensitive mantle edges. This functional 'conflict' can be overcome if the mantle (and valve) edges are modified into a zig-zag form: this reduces the distance between them without reducing the area of the gape. An appropriate form of zig-zag commissure would transform the gape into a zig-zag slit of uniform width (Fig. 61). Commissures with precisely this zig-zag form are found in some members of almost every major group of the articulates; they must have evolved very many times. They were formed during ontogeny by the modification of a wide variety of other deflections and crenulations of the valve

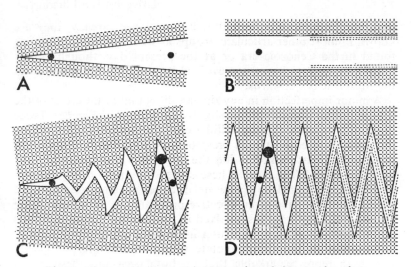

Fig. 61. Diagrams to illustrate interpretation of zig-zag valve edges as protective devices. A, B, straight valve edges, respectively perpendicular and parallel to the hinge axis, opened to a certain degree. C, D. corresponding zig-zag valve edges opened to same degree. Note that zig-zag slits greatly reduce maximum size of 'particles' admitted, without reducing area of aperture (somewhat larger 'particles' are admitted through the crests of the slits). Alternatively, as shown on right-hand slits of B and D, all parts of the zig-zag slit are within a short distance of the sensitive mantle edges (these two interpretations of protection are geometrically equivalent).[78,89,102]

edges, which also suggests their multiple origin (Fig. 62).[78] Even a single variety may have been evolved several times. For example, one distinctive kind of shell resulted from a sudden reduction in the 'wavelength' of the crenulation just before it was modified into a zig-zag form: on the valve surfaces a series of narrow radial costellae are replaced near the edge of the shell by a smaller number of broader costae (Fig. 62B). Zig-zags formed by this particular sequence of events occurred among the rhynchonellides in the middle Devonian period (*Nayunella*), then not again until the early Jurassic period about 200 million years later (*Rimirhynchia*), and then again for a third time in the early Cretaceous period (*Cyclothyris*) (Fig. 63). There is no trace of such shells in the intervening periods, and this is very unlikely to be due merely to the imperfection of the fossil record. Zig-zags were most abundant among the rhynchonellides from their earliest appearance in the Ordovician period until the Cretaceous period. But, as mentioned already, they are known in most other articulate groups as well. Curiously, none are found in the Cenozoic era or at the present day, though closely analogous zig-zags have been evolved by some of the oysters from the Mesozoic era onwards.

A minor imperfection in any zig-zag slit is that at the crests of the slit the mantle edges would be slightly further apart (Fig. 61). Twice in the history of the rhynchonellides even these points were provided for. One valve edge at each crest became modified into a slender spine, which was tucked inside the opposite valve edge when the shell was closed (Fig. 64A). These spines were presumably accompanied—and secreted—by portions of the sensitive mantle edge, and would therefore have given effective protection at the crests of the zig-zag slit. They were first evolved in the Silurian period, but then again about 100 million years later in the late Carboniferous period, in a species of the only punctate rhynchonellide (*Rhynchopora*). It is even more remarkable that on both occasions these spines appear to have evolved into a different and still more effective device (Fig. 63). In the Devonian period and then again much later in the Permian there were shells which passed through a growth stage with protected zig-zags; but in their final stages of growth the zig-zags died away and the spines became much longer (Fig. 64B). The gape thus returned to its original unmodified form; but it would now have been guarded by a grille of slender spines projecting at regular intervals from both valve edges, and presumably sheathed in sensitive mantle-edge tissue (Fig. 65). Somewhat similar grilles of internal spines can be seen on a few strophomenides (Fig. 41); and some highly aberrant members of that order (richthofeniids) transformed a grille of spines into a protective network or mesh (Fig. 87A).

Fig. 62. Zig-zag deflections in the commissures of fossil brachiopods, each in anterior and lateral views. Black pointers indicate geometrically (and functionally) 'ideal' point at which the deflection should 'die away' posteriorly; white points indicate actual positions: note generally close approximation between the two. A–D, Rhynchonellida: A, *Pugnax* (Carboniferous), B, *Nayunella* (Devonian), C, *Homoeorhynchia* (Jurassic), D, genus uncertain (Jurassic). E–F, Orthida: E, *Platystrophia* (Ordovician), F, *Parenteletes* (Permian). G, Spiriferida: *Crenispirifer* (Permian). H, Atrypida: *Hustedia* (Permian). I, Terebratulida: *Eudesia* (Jurassic).[78]

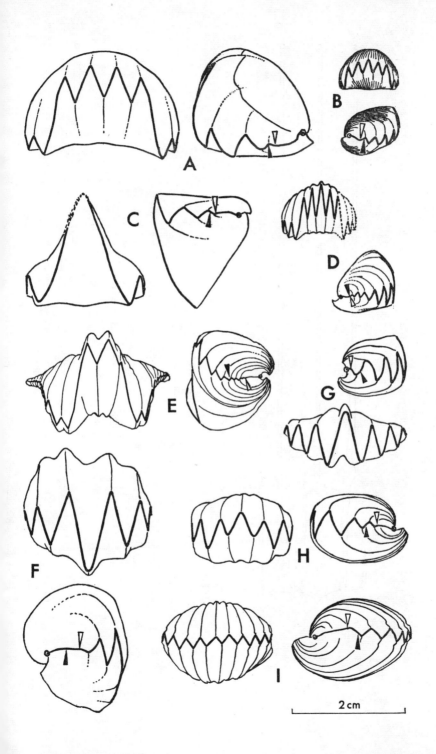

2 cm

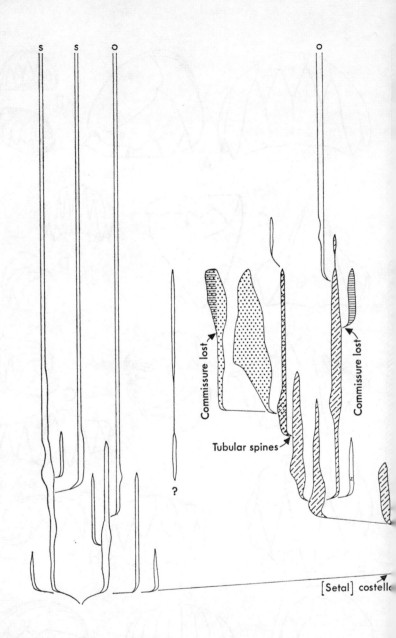

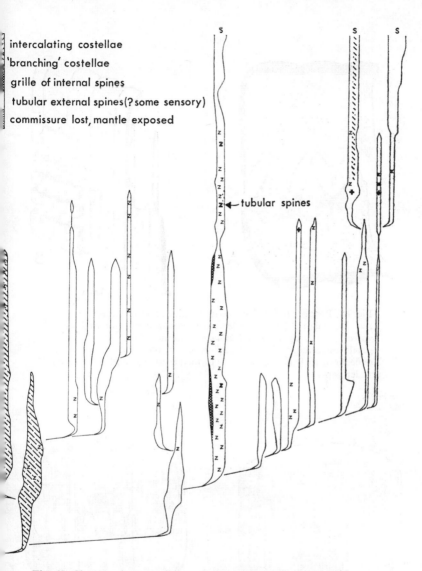

intercalating costellae
'branching' costellae
grille of internal spines
tubular external spines (?some sensory)
commissure lost, mantle exposed

←tubular spines

Fig. 63. Chart to show possible evolution of protective and associ-ated structures. Compare with Fig. 99; see also caption to Fig. 15. Z, genera with 'good' zig-zag commissures; (thicker Zs indicate genera with the particular type described on p. 111); S, setae present in extant genera; O, setae absent in extant genera; crosses, marginal projections of type described on pp. 110, 178 (see Fig. 98).[78,86]

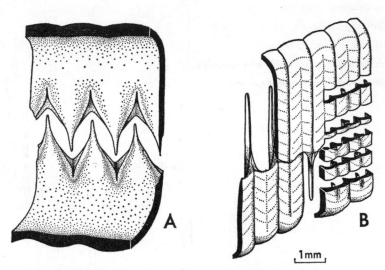

Fig. 64. Possibly protective apertural spines on the commissures of two Devonian rhynchonellides: A, *Obturamentella*, internal view, valves slightly open; B, *Glossinulus*, external view, valves closed, partly dissected.[102] In A the spines are at the crests of a zig-zag deflection; in B they could have formed a grille straddling a much wider gape.

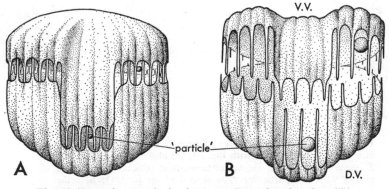

Fig. 65. Protective marginal spines on a Devonian rhynchonellide (*Uncinulus*): anterior views. A, at small degree of opening;[89] B, at wider degree of opening, giving less protection (see size of hypothetical 'particle') but much greater area for apertures.[86]

6

LOPHOPHORE AND FEEDING

The ciliary pump

The shell of a brachiopod closes for protection but must remain open for all metabolic interchange with the environment. Through the gape, guarded by sensitive mantle edges, the brachiopod obtains its food and oxygen and ejects its waste products. This traffic of materials is induced by the lophophore, which creates a continuous circulation of water currents—the *current system*—in and out of the mantle cavity.

The lophophore is attached to the anterior body wall at the back of the mantle cavity. From here it projects as a pair of long *brachia*. Each brachium consists of a cartilaginous *brachial axis*, bearing a row of long, slender flexible *filaments* (Fig. 66). The axis is variously looped or coiled, and is either attached to the dorsal mantle or, more commonly, suspended freely in the mantle cavity by various supporting structures. On the sides of each filament are rows of *lateral cilia*. The lashing of these cilia draws water through the narrow slits between the filaments, from the *frontal* to the *abfrontal* side (Fig. 67). This is the motive power for the current system. It is quite closely analogous to the ciliary pumps of many other invertebrates (e.g. bivalve and some gastropod molluscs, ascidians, etc.). Like them, its efficiency depends on the maintenance of a one-way circulation of water through an enclosed space. In a brachiopod, the enclosed space is the mantle cavity, and the one-way circulation is ensured by the way in which the filaments are arranged within it. However complicated the twisting or coiling of the brachia, they always divide the mantle cavity into an *inhalant* and an *exhalant chamber*. Water can only pass from one to the other by being pumped through the slits between the filaments (Fig. 68). The brachia also divide the gape between the valve edges into separate *inhalant* and *exhalant apertures*. Thus water is sucked through inhalant apertures into the inhalant chamber, pumped by the lateral cilia into the exhalant chamber, and ejected through exhalant apertures.[77]

This is very similar to the circulation through the mantle cavity of a bivalve mollusc. But there is an important difference. In brachiopods the filaments are never fused to one another or to any other structures. The 'partition' across the mantle cavity can only be maintained by the filaments being held in position by their own strength,

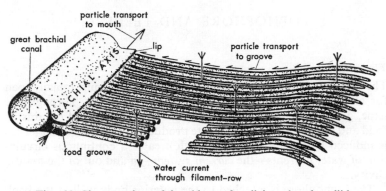

Fig. 66. Short section of brachium of a living rhynchonellide (*Notosaria*), showing structure and function of brachial axis and filament-row.[77]

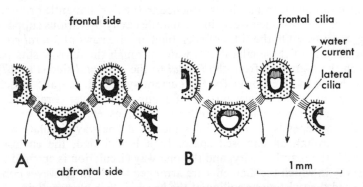

Fig. 67. Sections through the filament-row of (A) the inarticulate *Crania* and (B) the terebratulide *Platidia*, showing the water-pumping lateral cilia and particle-collecting frontal cilia. Epithelium stippled, connective tissue black, frontal muscle striated.[3,13,86]

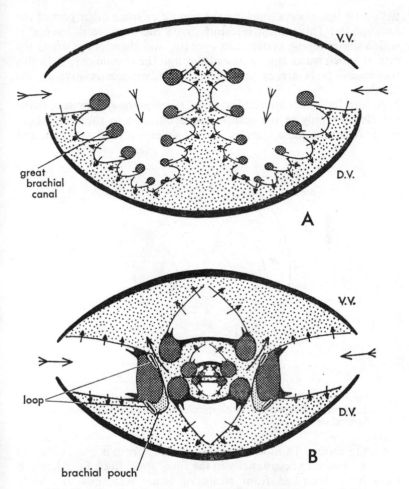

Fig. 68. Filtering current-systems of (A), spirolophe of a living rhynchonellide (*Notosaria*) and (B), plectolophe of a living terebratulide *(Waltonia,* Terebratellacea): transverse sections showing separation of exhalant chamber (irregularly stippled) from inhalant.[77] 'Tailed' arrows, unfiltered water; 'tail-less' arrows, filtered water; small arrows show direction of filtration between filaments. Note support of filaments by great brachial canal (heavy stippling) and, in B, partly by calcareous loop. Exhalant water emerges by median apertures which are not visible in these sections.

with their tips always touching the mantle or some other part of the lophophore. This almost certainly limits the pressure difference at which the pumping system can operate, and therefore restricts the rate at which water can be passed through the mantle cavity. In this the brachiopods are at an inherent disadvantage relative to the bivalves.

A few living terebratulides (e.g. *Terebratulina*) secrete calcitic spicules not only in the mantle tissue but also in the lophophore itself. Although these spicules never fuse into a rigid skeleton, and

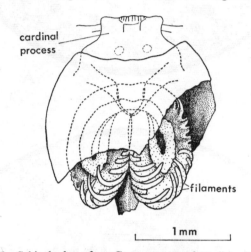

Fig. 69. Schizolophe of a Cretaceous terebratulide (*Meonia*; Terebratulacea), preserved with spicules showing form of filaments. Ventral view.[96]

are quite distinct from the brachidium, they form a mass coherent enough to survive occasionally in the fossil state. Such fossils, which have been described from strata of late Cretaceous and early Tertiary ages,[70, 96] are the only direct evidence of the form of the lophophore in fossil material (Fig. 69). They demonstrate that, at least in one group, the form of the lophophore and the size and spacing of the filaments have remained constant during the last 70 million years.

Feeding and rejection mechanisms

Like many other sessile invertebrates, brachiopods extract minute suspended particles of food from the sea-water. As the lateral cilia draw the water through the narrow gaps between the filaments, which after an early growth stage are arranged in two staggered rows, many suspended particles collide with the frontal surfaces of the filaments (Fig. 67). They are transported by *frontal cilia* to the base of the filaments. Here, on the frontal side of the brachial axis, is a ciliated *food groove* bounded by a muscular lip (Fig. 66). The particles travel along the food groove to the base of the brachia, where the groove leads directly into the mouth.

The activity of the lateral cilia is certainly under nervous control, and there is some evidence that the rate of filtration is geared to the concentration of food particles in the water, as it is in some bivalves. (Observations on living brachiopods can be highly misleading unless they are kept in aquaria with continuous circulation of fresh sea-water: in a closed circuit of filtered sea-water, for example, they may remain gaping open without any current-system operating, or more commonly will remain closed altogether.)

Normally, all particles that are small enough to remain in suspension are accepted indiscriminately, whether they have some food value (e.g. diatoms and dinoflagellates) or not (e.g. silt particles).[64, 77] There are no special sorting mechanisms; in this respect too the brachiopods are functionally inferior to the suspension-feeding molluscs. But if, for example, the water becomes highly turbid, a rejection mechanism comes into operation, and prevents the lophophore from becoming clogged up. The lip closes off the food groove and prevents any more particles from reaching the mouth, and the lateral cilia stop beating. Cells on the filaments secrete mucus, which enmeshes the particles. The frontal cilia reverse their direction of beat, and draw the rejected particles and mucus towards the tips of the filaments. (In *Crania* and the articulates there is no doubt that this is true ciliary reversal, though in *Lingula*[21] it is said to be due to adjacent tracts of cilia capable of beating in opposite directions, as in bivalve molluscs.) Eventually the mass of particles and mucus reaches the mantle, and is transported across its surface by mantle cilia. When it reaches the sensitive mantle edge the shell snaps shut. The consequent sudden outflow of water is generally sufficient to throw the mass of pseudofaeces clear of the shell. This rejection mechanism may affect only a tract of a few filaments, or a whole brachium, or the entire lophophore.[77] Reversal of the *lateral* cilia, and hence of the whole current system, has been reported in one species.[8] Other less radical rejection mechanisms are used to remove occasional larger particles (e.g. sand grains) without interrupting the normal filtering process.[21] All these rejection mechanisms are quite closely analogous to those used by bivalve molluscs.

Feeding and rejection mechanisms leave no trace in fossil brachiopods, but their uniformity in living species suggests that they have remained fairly constant during the history of the phylum.

Digestion and defaecation

From the mouth, the gut or digestive tube extends through the coelom, being held in position by mesenteries (Fig. 28). It comprises a narrow *oesophagus*, followed by a wider *stomach*, from which one or two pairs of ramifying digestive *diverticula* (the 'liver') branch off. Beyond the stomach the digestive tube continues as a narrow *intestine*. In living articulates the intestine opens by an *anus* (Fig. 70A), generally on the right lateral body wall—a rare departure from anatomical symmetry. In living articulates there is no anus and the intestine ends blindly, sometimes in an asymmetrical sac (Fig. 70B).

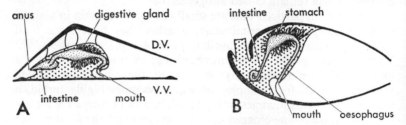

Fig. 70. Digestive systems of (A) a living inarticulate (*Crania*)[13, 24] and (B) a living rhynchonellide (*Hemithyris*),[49, 86] to show orientation within coelom (all other organs, and coelom, shown by regular stippling; lophophore omitted). Note anus in A and blind-ended intestine in B.

The presence or absence of an anus was formerly taken to be a difference of the highest importance, but it is now uncertain how fundamental it really is. The gut leaves no trace in fossil brachiopods, so we cannot tell whether the anus has always been characteristic of the inarticulates. In both classes the gut is blind-ended in the early growth stages (Fig. 91B), so that the opening of the anus could have been progressively delayed during the evolution of the articulates, or the change may have been paedomorphic. In any case the retention of a closed intestine is likely to have been connected with the evolution of the hinge, which permanently closed the posterior side of the gape.

The particles received from the food groove are moved down the digestive tube by peristaltic contractions of its walls. Cilia on its inner epithelium help to agitate the contents. The material is sucked in and out of the branches of the diverticula by rhythmic expansions and contractions of their muscular walls. Most of the actual digestion takes place in the diverticula, and appears to be intracellular.[2, 22, 24] The intestinal contents are rotated by ciliary action, and this is probably responsible for binding the undigested material into faecal pellets. In inarticulates these pellets are expelled from the anus; in articulates they are returned along the gut by anti-peristaltic contractions (the muscles in the oesophagal wall are striated) and are ejected from the mouth. In either case they are then transported by mantle cilia to the mantle edge, and are finally ejected (like the pseudofaeces) by a sudden snapping of the shell. As in other filter-feeding animals, the rejection of the faeces in the form of coherent pellets ensures that the faecal material is not passed again through the filtering system. Brachiopods, being sessile, have to rely on currents or scavengers to disperse the pellets, which would otherwise accumulate around the shell; but this is aided by the occasional swivelling of the shell around the pedicle.[77]

Living brachiopods snap their shells shut at fairly regular intervals for the ejection of faeces, and at other times for the ejection of pseudofaeces. The snapping reaction is immediately followed by the reopening of the shell, and the feeding is scarcely interrupted (Fig. 71).

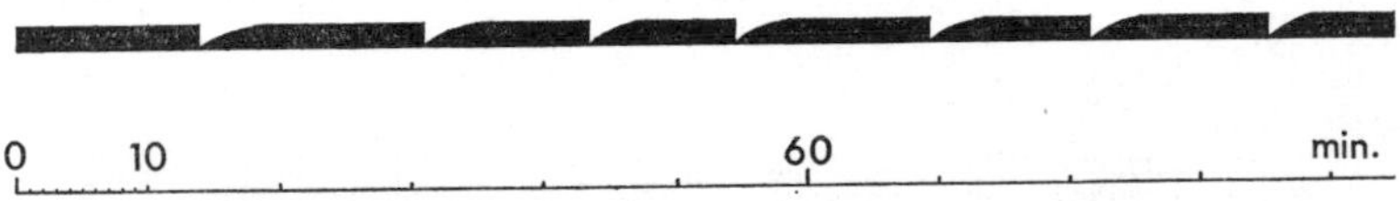

Fig. 71. Diagram to show periodic 'snapping' of a shell of a living terebratulide (*Neothyris*), with slower reopening.[77] Graph shows relative width of gape.

Excretory system

On either side of the coelom, supported by a mesentery, there is a small *nephridium* (strictly, a metanephridium). It consists of a trumpet-shaped tube lined with glandular and ciliated epithelium. The larger end opens into the coelom; the smaller end perforates the anterior body wall as a nephridiopore. The products of excretion are ingested by coelomocytes, drawn by ciliary currents into the nephridium, and eventually ejected through the nephridiopore enmeshed in mucus. Presumably they either leave the mantle cavity suspended in the exhalant current or else are transported by mantle cilia to the mantle edge and ejected there like the faeces and pseudofaeces.

Living rhynchonellides are unusual in possessing two pairs of nephridia. This was formerly interpreted as a last trace of the supposed metameric segmentation of the ancestral brachiopods; but since it is the only feature that can be claimed to be such a trace, it is more likely to be the result of a secondary doubling of the original pair of nephridia. No trace of the excretory system is preserved in the fossil state, so that interpretations of its history are wholly speculative.

Apertures

Apart from the lack of fusion of the filaments and the absence of sorting mechanisms, there is yet another way in which the feeding mechanisms of brachiopods are at an inherent disadvantage relative to the bivalve molluscs. The apertures by which water enters and leaves the mantle cavity are merely subdivisions of the general gape between the valve edges. As already mentioned, the mantle edges of brachiopods are never erected across the gape to separate one aperture from another, still less are they ever fused across the gape or produced into siphons. This debars the brachiopods from many of the modes of life that are exploited successfully by the bivalves. Only the aberrant inarticulate *Lingula* has to some extent overcome this limitation, by using erectile and mucus-covered setae to form structures equivalent to fused mantle edges and apertural siphons (Figs. 48, 72).

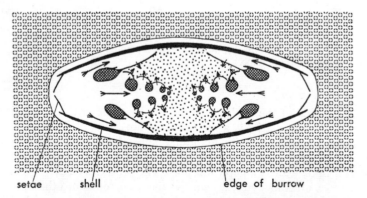

setae   shell   edge of burrow

Fig. 72. Filtering current-system of the burrowing inarticulate *Lingula*, shown by cross-section through spirolophe: note how the erectile setae complete the enclosure of the mantle cavity, like erectile or fused mantle edges in burrowing bivalve molluscs. Exhalant chamber stippled.[72]

In all other living brachiopods the apertures are separated by nothing more than the tips of some of the filaments, occasionally protected by a few setae (Fig. 56). In some, however, an anterior exhalant aperture is shifted dorsally or ventrally away from the inhalant apertures that flank it laterally, by the development of a median deflection in the commissure. The shift is not sufficient to isolate the apertures, but it does facilitate the outflow of water from the (high-pressure) exhalant chamber (Figs. 81E, 83E). The deflection is always in the direction in which the greater proportion of the pumping capacity (i.e. of the filaments) lies. Median deflections of this kind are very common among both living and fossil articulates; they give a valuable indication of the position of the apertures. (Although a median deflection usually generates a fold and sulcus during ontogeny (Fig. 7), observations on living species do not support the suggestion[101] that the fold and sulcus serve to channel *external* food-bearing currents: living articulates have only limited ability to swivel the shell, and do not take up a preferred orientation when placed in a steady current; primary functional significance must therefore be ascribed to the deflection itself, rather than to the fold and sulcus which it may—but does not necessarily—generate on the shell surface.)

In some extinct articulates the deflection deepened so much during ontogeny that in the adult shell it would have been capable of separating the apertures almost completely, at any normal angle of opening (Fig. 64). Such shells are found in many different groups, and these deep deflections must have evolved many times. A spatial separation of the inhalant and exhalant currents would be most important in a very quiet environment, where there might otherwise be a risk of recirculation. One group of terebratulides (e.g. *Pygope*), which lived in such environments during the Mesozoic era, developed isolated apertures to a remarkable degree. During ontogeny, and apparently also during phylogeny, an 'ordinary' ventral median deflection became deepened and isolated within an embayment of the commissure (Fig. 73A). This probably directed a jet of exhalant water upwards from the shell, far from either of the lateral inhalant apertures and away from the muddy substrate. Ultimately the lateral parts of the commissure met and fused across the median plane, and the exhalant aperture was left within a 'key-hole' in the shell (Fig. 73B).

Supporting structures

In living brachiopods there are four kinds of supporting structures for the lophophore. The brachial axes of inarticulates are held in

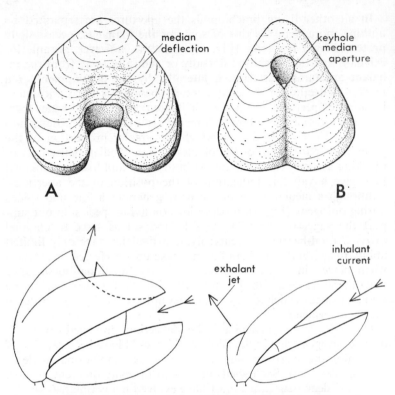

Fig. 73. Apertures in two aberrant Jurassic terebratulides ('*Pygope*' spp.; Terebratulacea); in A the commissure became strongly bilobed during ontogeny; in B the lobes actually fused and isolated the median deflection in a 'key-hole'. The lower figures suggest how such an increasingly isolated median aperture might have directed a jet of exhalant water away from the inhalant currents.[86,100]

position mainly by a mass of muscular and connective tissue. The position of the lophophore as a whole is controlled by lophophoral muscles, which are attached to the valves and run into the base of the lophophore at the front of the coelom. In fossil inarticulates there is no direct evidence of the course of the brachia.

In most of the living articulates the brachial axis is supported by a hydrostatic skeleton, a fluid-filled tube with thin muscular walls. The tube is the *great brachial canal* (Figs. 66, 68), which runs from the tip to the base of each brachium, but ends blindly at the base and is not connected with the coelom. (It also exists in inarticulates, but is relatively smaller.) As in the hydrostatic skeletons of other inverte-brates, there are muscle fibres in the wall of the canal, which act antagonistically against the incompressible fluid enclosed in the canal, and so maintain and control the position of the axis. A skeleton of this kind leaves no trace in the fossil state, but was pro-bably very common among the extinct articulates. The shells of a few fossil strophomenides show impressions on the inner surfaces of the valves, which appear to mark the form of a lophophore with some such 'soft' means of support.

The hydrostatic skeleton is commonly supplemented to some extent by a calcareous skeleton—the *brachidium*—which provides the main evidence of the lophophore in fossil brachiopods. The brachi-dium consists of one or more outgrowths of the secondary layer of the dorsal valve; and it is sheathed in, and secreted by, the same epithelium as the rest of the dorsal valve. (Calcite *spicules*, by con-trast, are formed by intracellular secretion within the connective tissue of the mantle and lophophore: they may become large and complex, but they never fuse into a solid skeleton.) The brachidium is invariably enclosed within the epithelia of the body wall. It may extend—within an extension of the body wall—roughly parallel to some part of the brachial axes; but is in no sense an extension into, still less a part of, the lophophore itself. There is no simple or neces-sary relation between the form of the brachidium and the form of the lophophore, and the reconstruction of the lophophore from the brachidium of a fossil brachiopod is therefore never a straightforward task.

The simplest form of brachidium is a pair of *crura*. These are calcareous spikes which project forwards from the hinge region of the dorsal valve, enclosed within the lateral body wall. They termin-ate against the walls of the great brachial canal at the points at which the brachia leave the body wall and extend freely into the mantle cavity. In this way the crura support and maintain the position of the *base* of the lophophore; the rest of the support comes from the hydrostatic skeleton. Thus in the fossil state the crura only identify the position of the base of the lophophore and give no indication of its form (Fig. 74). Crura grow by simple accretion at the anterior end. They are characteristic of all the rhynchonellides: even the earliest known species, from Middle Ordovician strata, have well developed crura, which may have existed already in a rudimentary form in their pentameride ancestors. Later pentamerides also possess well devel-oped crura, and may have evolved them independently (Fig. 88). Possible rudimentary crura are known in a few orthides. (The so-

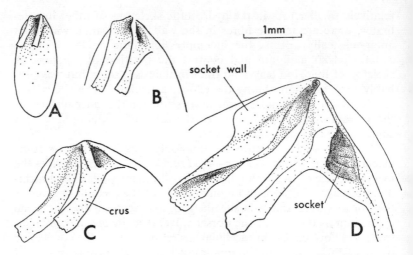

Fig. 74. Ontogeny of crura in a Cretaceous rhynchonellide (*Cretirhynchia*).[96]

called 'brachiophores' of orthides only rarely project far enough forwards to have functioned as possible supports for the base of the brachia: generally they seem to be nothing more than unusually massive inner walls of the hinge sockets.)

All further developments in the brachidium involved an important new property—the power of resorption in the epithelium sheathing the brachidium. There is no doubt that this resorption occurs during the growth of all the more complex brachidia in living brachiopods, and a simple geometrical analysis shows that it must also have occurred during the growth of all complex brachidia in fossils. Complex brachidia generally consist of a series of narrow ribbon-like *lamellae*, and growth-lines on these show clearly that they grow by accretion on one edge or surface and simultaneous resorption on the other (Fig. 75). During ontogeny these brachidia may undergo radical alterations in form, all traces of which are later destroyed by resorption: thus unlike most skeletal structures in brachiopods their development can only be pieced together by studying individuals of different ages. Brachidia of this kind are characteristic of the atrypides, spiriferides and terebratulides. The earliest, already showing evidence of resorption (Fig. 76), are found in the first atrypides in the Middle Ordovician, not long after the first crura-bearing rhynchonellides from which they had probably evolved. The brachidia of

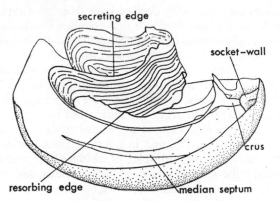

Fig. 75. Long-loop brachidium of a living terebratulide (*Dallina*; Terebratellacea),[7] showing resorption across growth-lines.

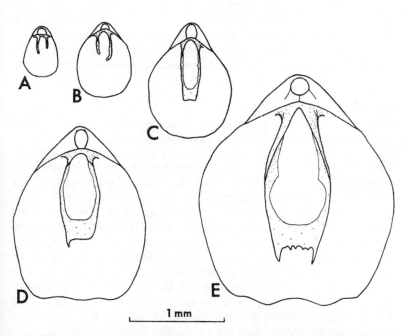

1 mm

Fig. 76. Ontogeny of brachidium in a primitive atrypide (*Protozyga*; Atrypacea),[110] showing development from simple crura (A, B), and implied resorption on inner edge of 'loop'.

terebratulides were probably derived later from an atrypide source. But there is considerable evidence that complex brachidia may have been evolved for a second time independently, in the Silurian period, producing—perhaps from orthide ancestors—the first spiriferides (spiriferaceans, cyrtiaceans). This would also represent an independent development of the power of resorption in the brachidial tissues (Fig. 88). Complex brachidia were originally developed as further shelly extensions from the crura; but in some of the later terebratulides (terebratellaceans) extensions from a median septum are also involved (Fig. 77), and in a few of these (kraussinids) the crural part has been lost. Complex brachidia also occur in isolated genera within the strophomenides, pentamerides and orthides:[110] these almost certainly represent separate, but relatively 'unsuccessful', developments of the power of resorption in the growth of the brachidium.

The fourth and last mode of support of the brachial axes is the only one which commonly preserves direct evidence of the form of the lophophore in the fossil state. The axes are fused throughout to the dorsal mantle, and rest in *brachial grooves* in the inner surface of the dorsal valve, bounded on one or both sides by narrow ridges (Fig. 78). This direct support is known only in one small extant group (thecideaceans), but there are similar grooves in one of the most aberrant extinct strophomenide groups (lyttoniaceans).

Growth of the lophophore

The lophophore first appears as a group of a few pairs of ciliated projections on the body of the larval brachiopod. These grow into the first pairs of filaments. New filaments are subsequently added in pairs, always at the distal ends of the brachial axes and never by intercalation elsewhere.[5] Although the number and length of the filaments thus increase progressively, the width of the slits between them remains almost constant—probably at the optimum width for the ciliary action. The pumping mechanism itself also remains unchanged throughout the growth of the brachiopod. At every stage the cilia pump water from the frontal to the abfrontal side of the filaments, and the brachia divide the mantle cavity into separate chambers with separate apertures to the exterior. There is every reason to suppose that these features, which are essential to the functioning of a ciliary pump of this kind, were as uniform among extinct brachiopods as they are among the living species.

As a brachiopod grows, its metabolic requirements increase roughly in proportion to the volume of its body. On the other hand

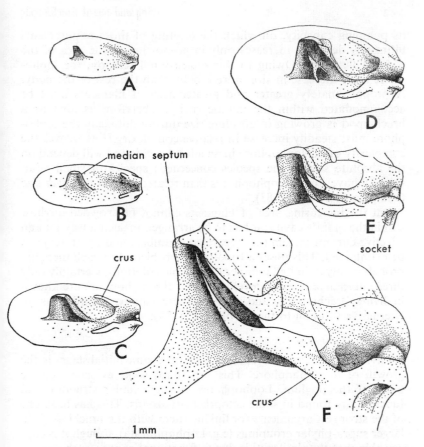

Fig. 77. Ontogeny of long-loop in a Cretaceous terebratulide (*Magas*; Terebratellacea), showing important role of median septum.[96]

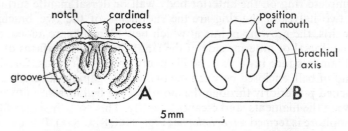

Fig. 78. Ptycholophe of a living strophomenide (*Lacazella*; Thecideacea): A, looped grooves on interior of dorsal valve; B, course of brachial axis (filaments not shown).[57]

its pumping capacity, on which the meeting of those requirements ultimately depends, increases only in proportion to the area of the rows of filaments. Owing to this dimensional relation, the lophophore must increase in size more rapidly than the rest of the body. A proportionately greater and greater area of filaments must be accommodated within the mantle cavity. Therefore as long as a brachiopod is growing in absolute size during ontogeny the lophophore must steadily increase in relative complexity.[77] Moreover the complexity of the lophophore in an adult brachiopod will depend on the absolute size of the species concerned: species of small adult size will have simpler lophophores than related species of larger size (compare Figs. 80 and 83).

But the increasing area of filaments cannot be crowded anyhow within the mantle cavity: it has to be arranged in such a way that an effective current system (i.e. separated chambers and apertures) can be maintained. This poses a topological problem, to which there are probably only a limited number of possible solutions. Certainly only three alternative lines of development of the lophophore are known in living brachiopods, and each of these can be regarded as an alternative solution to the problem (Fig. 79).

Simple lophophores

All living species, however, pass through the same initial stages in the growth of the lophophore. This is perhaps some evidence for the unity of the phylum. Lophophores of very similar structure and function are found in some ectoproct bryozoans. This has been one of the strongest arguments for linking them with the brachiopods in larger supra-phylar groupings (e.g. Lophophorata), though it is conceivable that the similarity is due to functional convergence.

In the earliest stages of growth, the filaments are arranged in an incomplete ring on the anterior body wall or dorsal mantle surface. The two halves of the ring are the rudiments of the two brachia; their tips, the growing points at which new filaments are added, are on the dorsal or anterior side of the ring; the frontal surfaces of all the filaments face inwards. The filaments project forwards, forming a kind of bell, enclosed between the two widely gaping valves. Water is sucked posteriorly through the mouth of the bell, drawn outwards between the filaments, and escapes laterally. This growth-stage of the lophophore is termed a *trocholophe* (Figs. 80A, 81A, 83A). It is universal in the small and early growth stages of all living brachiopods, and also characterises the adults of one genus of exceptionally small size (*Gwynia*).

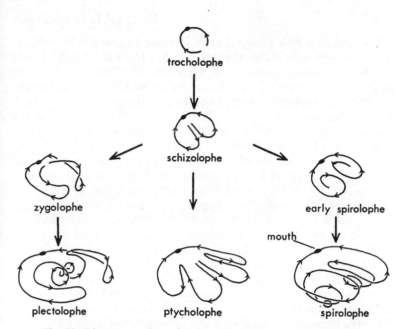

Fig. 79. Diagram to show the three alternative ontogenetic 'path-
ways' in the growth of the lophophore in living brachiopods.[77]
Each drawing shows a perspective view of the course of the brachial
axis (filaments omitted). Arrows on the axis indicate the direction of
transport of food particles towards the mouth.

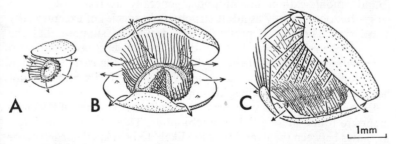

Fig. 80. Ontogeny of lophophore and its current system in a living
terebratulide of small size (*Pumilus*): the lophophore does not
develop beyond the schizolophe condition (B, C). Setae shown only
in C. Ventral valve above.[77,86]

The first sign of a proportionate increase in complexity appears in the next growth stage, the *schizolophe*. Here the brachial axes have buckled into a bilobed form. They are still adherent to the body wall and dorsal mantle; the valves still gape widely; and most of the filaments still project forwards like a bell, drawing water in from the front and expelling it laterally. But in the median indentation the filaments arch over to form a kind of tunnel, which receives the water pumped by these filaments and passes it anteriorly. Like the trocholophe, the schizolophe is almost universal in the small and early growth stages of living brachiopods (Figs. 81B, 83B), but in a few species, which become adult at a very small size, it is also the final form of the lophophore (Fig. 80B, C). Such species are found among the lingulides (*Pelagodiscus*), strophomenides (*Thecidellina*) and the terebratulides (*Argyrotheca*; *Pumilus*); some of these probably evolved independently by neoteny from related brachiopods with more complex lophophores.

In the schizolophous growth stage the first rudiments of the various supporting structures become apparent. The great brachial canal is sealed off from the coelom and becomes effective as a hydrostatic skeleton. In the rhynchonellides and some terebratulides (terebratulaceans), crura grow forwards and support the posterior side of the schizolophe. In other terebratulides (terebratellaceans), there is also a median septum which grows from the floor of the dorsal valve and supports the median indentation of the schizolophe (Fig. 77A, B). A pair of lamellae may extend from the crura parallel to the brachial axes and join up with the septum. In living strophomenides, on the other hand, the axes are supported throughout in brachial grooves on the dorsal valve. Structures similar to all these examples can be found among extinct brachiopods, generally as transient growth stages but occasionally as adult structures in species of exceptionally small size. There is at present every reason to suppose that the trocholophe-schizolophe line of development has been universal throughout the history of the phylum. Many of the early Cambrian brachiopods (e.g. acrotretaceans) are very small in size and may not have advanced beyond the schizolophous stage.

Beyond this point in ontogeny, three alternative lines of development are found among living brachiopods. There are no true intermediates between them, and they are likely to be related to each other only through small neotenous forms with schizolophes.

Spiral lophophores

In a schizolophe the brachial axes are fused throughout to the body wall and dorsal mantle. The first and perhaps most important line of development escapes from this limitation and allows the brachia to be coiled freely in three dimensions within the mantle cavity. When a schizolophe develops into a *spirolophe*, the tips of the brachia diverge from one another and from the mantle surface, until eventually the brachia grow into a pair of spirals. The tips of the filaments touch either the mantle or another part of the spiral, so that the whole spirolophe divides the mantle cavity into separate chambers (Figs. 68A, 72). The interior of the spirals forms part of either the inhalant or the exhalant chamber, depending on the direction in which the brachia are coiled. The inhalant spirolophe is the more common among living brachiopods but the exhalant is also known (*Discinisca*).[63] The arrangement of apertures around the gape is gradually transformed during ontogeny. The median inhalant aperture of the schizolophe, flanked by exhalant openings, is gradually replaced by an exactly reverse arrangement, with a median exhalant aperture. Very commonly this aperture becomes marked by a median deflection (Fig. 81).

Spirolophes are characteristic of all the living inarticulates and rhynchonellides. They are supported principally by a hydrostatic skeleton or other 'soft' supporting structures, which leave no trace in fossil material. But the wide distribution of spirolophes among living brachiopods suggests that this has been a very common form of lophophore throughout the history of the phylum. It may have been the first way in which the lophophore grew beyond the schizolophous stage, and thereby enabled brachiopods to become moderately large in size. Clear impressions of spirolophes have been found on the thickened valves of several Palaeozoic strophomenides, which would otherwise betray no sign of the form or even existence of the lophophore (Fig. 88); and it is reasonable to assume that their relatives had similar lophophores, even though no impressions of them are preserved. (Many of these strophomenides have looped ridges, the so-called 'brachial ridges', on the inner surface of the dorsal valve: these have been interpreted as the attachments of schizolophes or partially supported spirolophes,[109] but they are quite unlike any lophophoral supporting structures in living brachiopods, and their function must at present be regarded as highly uncertain.)

The spirolophes of living rhynchonellides are supported basally by crura. As already mentioned, similar crura are characteristic of this order from its origin in the Ordovician period, and also of the later pentamerides. All these crura are likely to have supported spirolophes, although there is no direct evidence of this.

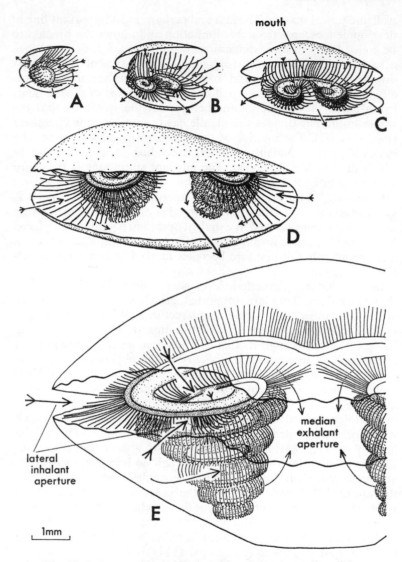

mouth

median exhalant aperture

lateral inhalant aperture

1mm

E

Fig. 81. Ontogeny of lophophore and its current system in a living rhynchonellide (*Notosaria*): A, trocholophe; B, schizolophe; C–E, spirolophes; 'tailed' arrows, unfiltered (inhalant) water; 'tail-less' arrows, filtered (exhalant) water; (valves in adult (E) are drawn as though transparent, to show full size of spiral brachia).[77] Setae omitted (see Fig. 57). Ventral valve above.

Spirolophes were also probably characteristic of the atrypides and spiriferides. All members of these orders have brachidia consisting of a pair of spiral lamellae, the *spiralia*, extending from a pair of crura. As already mentioned, the growth of such brachidia certainly involved the power of resorption in the epithelia sheathing the lamellae. Both orders are extinct, so that the reconstruction of their lophophores is not without uncertainty. But the form of the spiral lamellae of some of them (atrypaceans) is virtually identical to the course of the brachial axes in living rhynchonellides. The simplest and most plausible reconstruction is to assume that the lamellae ran parallel to the brachial axes and that the lophophore was a spirolophe. (Against this interpretation it has been argued that each lamella supported a double loop of the brachial axis, i.e. a double row of filaments, forming a 'deuterolophe'.[110] This would be unlike the lophophore of any living brachiopod. Moreover this reconstruction depends on the assumption, which is demonstrably incorrect for living brachiopods, that the brachial axes were bound to run parallel to the edges of the brachidial lamellae throughout ontogeny. It also poses insoluble difficulties for the current system, for it seems to be topologically impossible to arrange the filaments of a 'deuterolophe' in any way that would give an effective current system comparable to that of any living brachiopod.)

On the first interpretation it is possible to reconstruct the filaments in the only orientation that would have enabled the spirolophe to create an effective current system.[72] According to the direction of coiling of the spiralia, some of the atrypides (some atrypaceans) would have had inhalant spirolophes like those of living rhynchonellides (Fig. 82A), whereas most of the others, and all the spiriferides, would have had exhalant spirolophes like that of the living *Discinisca* (Fig. 82B). Neither type has any obvious inherent advantage over the other; in fact the 'exhalant' ones were much more abundant than the 'inhalant', and outlasted them by about 200 million years, although it is the inhalant type that is more common among the spirolophes of living brachiopods. Spiralia are usually 'moulded' quite accurately to the shape of the shell-cavity, especially in the later genera. This would have given the maximum area of filtering filaments possible within the space of the mantle cavity. In atrypides and spiriferides, as in living brachiopods, both types of spirolophe would have had lateral inhalant apertures flanking a median exhalant aperture; the latter is very commonly marked by a median deflection.

By homology with the brachidia of living brachiopods the spiral

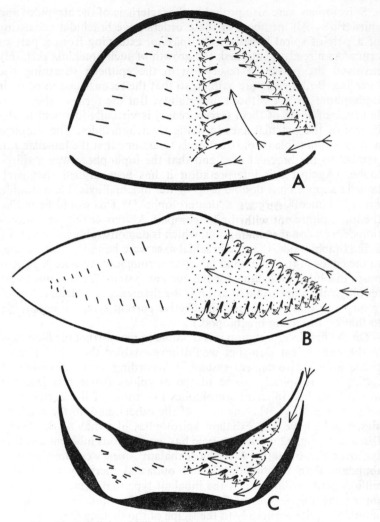

Fig. 82. Spiral brachidia, and inferred spirolophes and filtering current systems of, A, a Devonian atrypide (*Atrypa*), B, a Carboniferous spiriferide (*Spirifer*), and C, a Triassic strophomenide (*Koninckina*). Cross-sections showing, on left, spiral brachidium as preserved; on right, reconstruction of spirolophe and water currents (exhalant chamber stippled, inhalant currents shown by 'tailed' arrows). Dorsal valve above.[72]

lamellae must have been enclosed in long extensions of the body wall. Possibly the development of this complete calcareous support was accompanied by the reduction or even the atrophy of its functional equivalent, the hydrostatic skeleton. If so, the support for each whorl of the brachia could have been relatively slender. This could explain why the successive whorls are often more closely spaced than those in living brachiopods; and closely spaced whorls with slender supporting structures would have enabled a relatively larger area of filtering filaments to be borne on a spiral brachium of given size. Against this inherent advantage of spiral brachidia would be the disadvantage that they were evidently extremely delicate structures (they were often broken before burial, even in otherwise finely preserved specimens). Possibly this has contributed to the relative lack of 'success' of spiral-bearing groups, which have all been extinct since early Jurassic time, relative to the rhynchonellides with their more robust method of support for the spirolophe.

The brachidia of many atrypides developed further skeletal structures. A lamella—the *jugum*—linking the spiralia near their base, probably helped to support the base of the spirolophe or some part of the body wall. In two groups (atrypaceans, athyridaceans) the brachidium seems sometimes to have been only loosely connected with the crura; in such shells the jugum may have been the seat of muscles serving to rotate the whole lophophore into the most effective position when the shell opened.[31] In one of these groups (athyridaceans) the jugum became extremely complex. It developed a forked extension which curved round parallel to the bases of the spiralia. Occasionally this was even extended into a pair of complete secondary spirals running parallel to the main spirals throughout their length. This extraordinarily complex structure was evolved during the Devonian period (*Kayseria*) and again nearly 200 million years later in the Triassic period (*Diplospirella*, etc.). Possibly it provided further support for the brachial axes.[72]

As already mentioned, spiral brachidia first appeared in Middle Ordovician atrypides (atrypaceans), probably from rhynchonellide ancestors; a few members of this group (leptocoeliids) seem later to have lost their calcareous support for the lophophore. The spiriferides may have evolved spiralia independently, perhaps from orthide ancestors, early in the Silurian period (Fig. 88). A spiral brachidium was also evolved, quite independently, by one Triassic strophomenide (*Thecospira*). Significantly, its gutter-shaped lamellae are unlike those of any spiriferide, although it probably supported a very similar spirolophe.[84] Another strophomenide group (koninckinaceans) evolved *double* spiral brachidia, like those of the atrypides mentioned above, at about the same time (Fig. 82c).

Terebratulide lophophores

The second of the alternative lines of development from the schizolophe also allows the brachia to be coiled in three dimensions. But in addition, the fusion of certain parts of the supporting structures allows the available space in the mantle cavity to be used even more effectively. The first stage involves a lateral twisting of the lobes of the schizolophe. This results gradually in the reversal of the original arrangement of the apertures, as in the growth of a spirolophe. These lateral lobes then grow forwards off the dorsal mantle surface and project freely into the mantle cavity; but the brachial axes remain united across the floor of each lobe, so that the current system continues to be effective. The uniting of the axes is due to the fusion of the (otherwise doubled) great brachial canal into a single supporting tube (Fig. 68b). This is the first novel feature which makes this line of development possible.

This stage of growth is termed a *zygolophe*. Among living brachiopods it is confined to the terebratulides, and is never more than a transient stage (Fig. 83c). It is always supported by some form of brachidium. In one group (terebratulaceans) this consists of lamellae extending from the crura and later uniting on the dorsal side of the lophophore to form a short loop. This supports the base of the zygolophe where it is attached to the body wall. In most members of the other extant group (terebratellaceans) the zygolophe is supported partly by lamellae extending from the crura and partly by some projecting from the crest of the median septum. In one small group (kraussinids) the support is only of the latter kind. All these brachidia grow by simultaneous secretion and resorption; all the lamellae are enclosed in extensions of the body wall. Some of the earliest terebratulides (stringocephalaceans) had loops which may have supported zygolophes even in the adult stage. There is no direct evidence to confirm this, for these loops have no close analogy among living brachiopods; but the most plausible explanation for their form is that they supported the lateral lobes of a zygolophe, perhaps fused back to back across the median plane (Fig. 84).

The first development of zygolophes can probably be located at the origin of the terebratulides themselves, early in the Devonian period. The loops of these early terebratulides and the spiral brachidia of contemporary spiriferides have little in common in their adult forms, but their early growth-stages are quite similar. This suggests

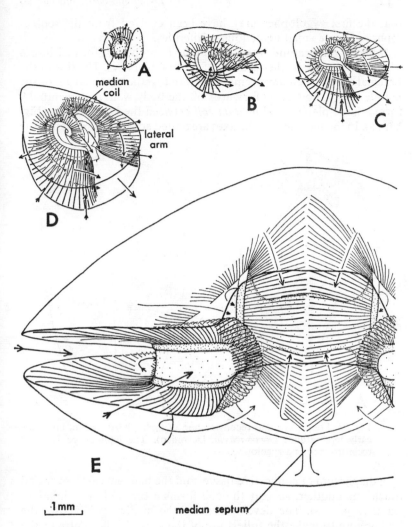

Fig. 83. Ontogeny of plectolophe in a living terebratulide (*Waltonia;* Terebratellacea): A, trocholophe; B, schizolophe; C, zygolophe; D, E, plectolophe (all except A drawn as though valves were transparent). Ventral valve above.[77]

that the first zygolophes may have been evolved from the schizo-lophes of some small neotenous spiriferides.

In living terebratulides the zygolophe is soon transformed into a *plectolophe* by the further growth of the brachial axes. The tips of the brachial axes, which have remained close together at the back of the mantle cavity, now grow forwards off the body wall, and eventually form a large plane spiral *median coil* between the lateral lobes (Fig. 83D, E). In the median coil the axes are united across the median plane

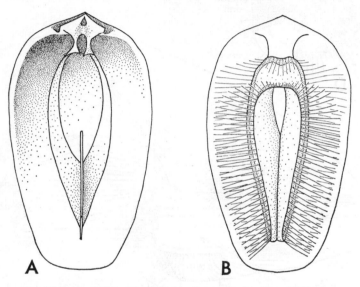

Fig. 84. Loop (A)[95] and reconstruction of lophophore (B)[86] of an early terebratulide (*Renssaeleria*, Devonian). The lophophore is reconstructed as a zygolophe.

by a diaphragm of connective tissue, and the filaments arch over and touch one another, so that the coil forms a tunnel full of inhalant water (Fig. 68B). The development of the median coil allows the lophophore to make the fullest use of the space in the centre of the mantle cavity (Fig. 83E); but it is dependent on the existence of the diaphragm, without which the current system could not be effective.

The plectolophes of living terebratulides are supported by a variety of brachidia, which are developments of those of the zygolophous stage. The base of the plectolophe may be supported by a *short loop* (terebratulaceans) (Fig. 28). Alternatively, by a complex meta-morphosis, the lamellae connected to the crura and to the septum may link up with each other (Fig. 77) and then both become more or less detached from the septum, forming a *long loop* (terebratellaceans) (Fig. 75). In this the lamellae run roughly parallel to the brachial axes in the lateral lobes, but are still enclosed within extensions of the body wall, accompanied by extensions of the coelom, the 'brachial pouches' (Fig. 68B). Long loops only support the base of the median coil, where it is attached to the body wall, and never extend into the median coil itself. Similar long loops in some extinct terebratulides (zeilleriaceans) seem to have developed solely from the crura, without the intervention of a median septum, and may have evolved independently. But most of these long loops in fossil terebratulides would have allowed room for a median coil, and therefore probably supported plectolophes, although as usual there is no direct evidence to confirm this. Loops analogous to those known to support plectolophes in living species first appeared among the earliest terebratulides in the Devonian period. This may mark the first acquisition of the diaphragm that made a median coil and therefore a plectolophe an effective possibility. Certainly the plectolophe is by far the most abundant and 'successful' form of lophophore among living brachiopods.

Lobate lophophores

The third and last alternative mode of development of the lophophore seems, by contrast with the others, inherently less effective. It involves nothing more than the expansion of a bilobed schizolophe into a multi-lobed *ptycholophe*. Further indentations of the brachial axes are added laterally, and their filaments form additional exhalant tunnels; otherwise the current system remains much the same. Only two living genera (*Lacazella*, *Megathiris*) have ptycholophes; in both of them the lophophore never develops beyond a four-lobed stage (Figs. 78, 85A), and the adult shells are only slightly larger than related schizolophous species. But one of them (*Lacazella*) has fossil relatives (thecideaceans) which developed more complex ptycholophes with up to twenty lobes, and correspondingly grew to a larger adult size (Fig. 85B). In this group the brachial grooves enable the lophophore to be reconstructed with virtual certainty. These complex ptycholophes seem to have been evolved during three periods in the history of the group, first in the late Triassic, again but in a different form in the early Jurassic, and again much later in the Cretaceous (Fig. 97). But even the largest of these shells is still fairly small, by comparison with contemporary brachiopods with spirolophes or plectolophes. This may reflect a limitation inherent in any normal ptycho-

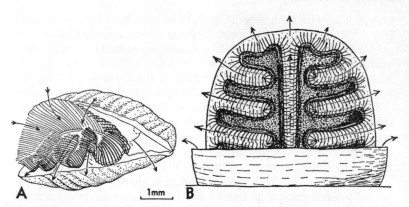

Fig. 85. Current systems of ptycholophes: A, simple ptycholophe of a living terebratulide (*Megathiris*, Terebratellacea) seen in anterolateral view;[8] B, reconstruction of complex ptycholophe of a Triassic strophomenide (*Bactrynium*, Thecideacea) seen in anterior view.[84] Note unusually wide opening of shell, and ill-defined apertures.

lophe. For however buckled the brachial axes become, they must always remain attached to the mantle surface: otherwise the inhalant interior would not be separated from the exhalant water outside, and the whole current system would be ineffective. The looping of the brachia must therefore be confined to the plane of the dorsal mantle, and cannot take full advantage of the three-dimensional space of the mantle cavity.

There is, however, one extinct group of strophomenides (lyttonia-ceans) in which ptycholophes seem to have escaped from this inherent size-limitation. These shells have looped grooves of up to forty or more lobes around the edges of a similarly lobed dorsal valve. Their ancestry is highly uncertain. The earliest forms (*Poikilosakos*), which lived in late Carboniferous time, were fairly small and had few lobes; the later members passed through similar stages during ontogeny but became much larger and more complex. Members of this group were abundant and varied during the Permian period, but became com-pletely extinct at its end. These brachiopods were certainly aberrant, as mentioned earlier, in having abandoned the normal protective function of the shell, and the dorsal valve seems to have been adapted purely into a supporting structure for the elaborate lophophore. But by maintaining a series of slots in the edge of the valve, between the lobes of the lophophore, water filtered into the exhalant 'tunnels' could have passed directly 'through' the dorsal valve (Fig. 86). Certainly these aberrant ptycholophes grew on a much larger scale than any others: even the smallest and simplest are much larger than normal ptycholophes of comparable form.[85,121]

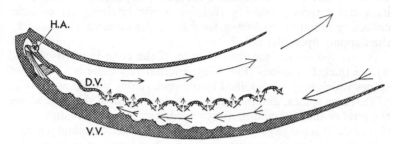

Fig. 86. Reconstruction of possible current system in an aberrant Permian strophomenide (*'Lyttonia'*, Lyttoniacea): note water being filtered by filaments fringing the slots in the dorsal valve.[87]

Rhythmic feeding mechanisms

The basic ciliary feeding mechanisms of living brachiopods, whether inarticulate or articulate, are remarkably uniform; and if the lophophores and current systems of most fossil brachiopods are reconstructed on the assumption that the same basic mechanisms were operative, the preserved morphology can be interpreted functionally in a consistent and coherent way, as outlined in previous sections of this chapter.

There are, however, a few fossil brachiopods which are so abnormal in their morphology that any such analogy with living species is likely to be misleading. The chief examples of such brachiopods are found among the astonishingly rich and varied strophomenide faunas of the Permian period; and the most striking of these belong to a group (richthofeniid strophalosiaceans) which has already been mentioned for its highly aberrant morphology.

In these shells the ventral valve is modified into a deep cone, and the dorsal valve forms a thin operculum recessed within the cone (Fig. 87). There is no preserved trace of a lophophore; if it existed at all it would clearly have been situated below the operculum. But if it operated with the normal ciliary mechanisms, no conceivable current system could have been effective from so recessed a position. This alone would suggest that the feeding may have been abnormal. Moreover, the curiously elegant internal hinge mechanism (Fig. 18), and the paradox of a powerful musculature and very lightly constructed operculum, suggest that the operculum habitually flapped vigorously up and down. Model reconstructions prove that such movements would have created powerful currents and eddies in and out of the interior of the cone. Thus the lophophore might have lost its usual current-producing function while retaining its particle-collecting function, gathering food from the currents produced by the flapping operculum.[73, 87]

Above the operculum the aperture of the cone is often covered with a thicket of spines (Fig. 87B); alternatively there is a screen of spines which may be modified into a remarkably perfect mesh (Fig. 87A). These spines, and indeed the whole interior of the cone above the level of the operculum, must have been covered in mantle tissue (Fig. 5B). This suggests that they may have formed a second particle-collecting device, the mucus-secreting ciliated mantle surface being adapted from the usual purely rejecting function to a function of capturing food particles and transporting them to the mouth. In species with a mesh this was clearly combined with a function of protecting the interior from harmfully large particles.

The entire morphology of these curious brachiopods is intelligible in terms of such a rhythmic feeding mechanism. Although unparalleled in living brachiopods, such a mechanism would have been somewhatanalogous to that of the septibranch bivalve molluscs. Similarly, it might have enabled these brachiopods to capture and feed on a wider range of food than was available to their ciliary-feeding ancestors. It is possible to envisage the gradual evolution of such a mechanism, from the normal ciliary feeding mechanisms of brachiopods, by a gradual adaptation of the snapping reaction of the shell from a rejecting into a feeding device.

Stages in this process may be represented among many other members of the strophomenides from which these most aberrant brachiopods were derived, for their morphology suggests that they could already have operated a rhythmic feeding mechanism at least

Fig. 87. Reconstruction of possible rhythmic feeding mechanisms in two highly aberrant Permian strophomenides ('*Richthofenia*' spp., Strophalosiacea). The left-hand figures show the thin trap-door-like dorsal valve half-open; the right-hand figures, half-closed: note the currents and eddies that result from rapid movements of the valve. In A, a fine shelly mesh would have screened large particles from entering; in B, thickets of spines would have intercepted the strongest currents and may have served to collect small food particles. Lower parts of conical ventral valves omitted.[87]

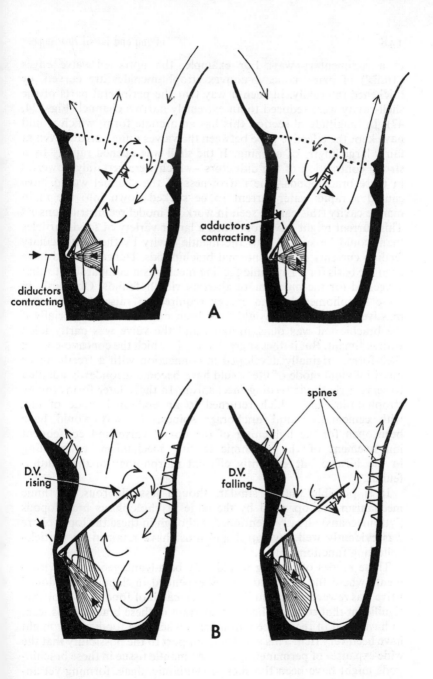

adductors
contracting

diductors
contracting

A

spines

D.V.
rising

D.V.
falling

B

in a rudimentary way. For example, the upturned valve edges ('trails') of many concavo-convex strophomenides are curved, or thickened internally, in such a way that the peripheral parts of the shell-cavity were reduced to an extremely narrow channel (Figs. 24, 47). In longitudinal section this has an arcuate form, which would have kept the effective gape between the valves quite narrow, even at fairly large angles of opening. If the shell was opened rapidly by a strong contraction of the diductors—which were certainly powerful in these brachiopods—the narrowness of the channel would have caused a rapid water-current to be sucked down into the main mantle cavity (this can be seen in working model reconstructions).[86] This current might have contained a larger variety of food-particles than would be sucked into the mantle cavity by the gentle ciliary feeding currents of more normal brachiopods. Hence it could have been the basis for a rhythmic feeding mechanism comparable to that suggested for the much more aberrant richthofeniids. Obviously in these strophomenides the energy required to raise the relatively massive concave valve would have been much greater, especially if the brachiopod was quasi-infaunal and the valve was partly filled with sediment. But it does suggest a way in which the concavo-convex shell-form, originally developed in connection with a free-lying or quasi-infaunal mode of life, could have become secondarily adapted to serve a rhythmic feeding mechanism. In those later forms (many strophalosiaceans) which returned to an epifaunal mode of life, with cementation and anchorage spines, the way would have been open for the lightening of the dorsal valve and consequent improvement of the rhythmic feeding mechanism, culminating in the bizarre but evidently efficient morphology of the richthofeniids.

It is possible that a similar, though less vigorous, rhythmic mechanism was operated by the large ptycholophous brachiopods (lyttoniaceans) already mentioned, although in these the lophophore was evidently well developed and must have retained its particle-collecting functions intact.[87]

These modes of feeding would only be advantageous in environments where the muscular energy expended in flapping the dorsal valve was rewarded by a sufficiently rich catch of food. It is probably significant that some of the most aberrant of these brachiopods seem to have lived in reef-like environments where the food supply might have been very rich. This also lends support to the possibility that the wide expanses of permanently exposed mantle tissue in these brachiopods might have been the sites of symbiotic algae, forming yet another supplementary source of food: certainly there is a curiously close analogy to the tridacnid molluscs of modern coral reefs, with their symbiotic zooxanthellae in similarly exposed areas of mantle tissue.[118]

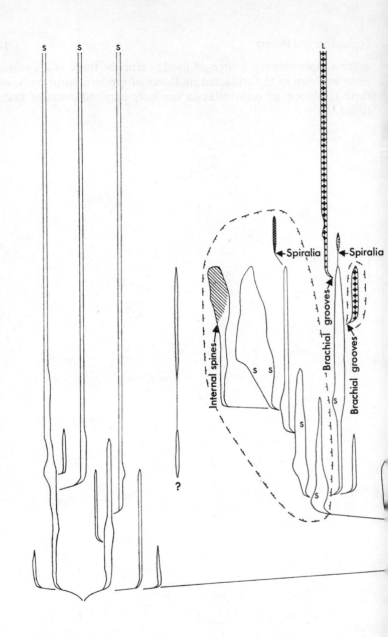

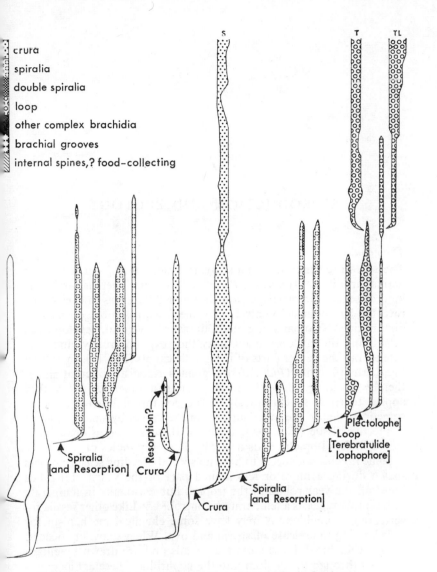

crura

spiralia

double spiralia

loop

other complex brachidia

brachial grooves

internal spines,? food-collecting

Spiralia
[and Resorption]

Resorption?

Crura

Crura

Spiralia
[and Resorption]

[Plectolophe]
Loop
[Terebratulide
lophophore]

Fig. 88. Chart to show possible evolution of brachidia, lophophores and feeding mechanisms. Compare with Fig. 99; see also caption to Fig. 15. L, extant genera with lobate lophophore (e.g. ptycholophe); T, extant genera with 'terebratulide' lophophore (e.g. plectolophe); S, extant genera with spirolophe, and fossil genera with spiral impressions; dashed line surrounds groups with possible rhythmic feeding.[86]

7

REPRODUCTION AND ECOLOGY

The gonads

The sexes are separate in most living brachiopods, but a few are hermaphrodite (*Argyrotheca, Pumilus, Platidia*).[2] There is rarely any kind of sexual dimorphism, though minor differences in shell-form, revealed by statistical analysis of some fossil assemblages, may be dimorphic.[98] Two pairs of gonads, dorsal and ventral, develop from cells lining the coelom; at maturity they expand greatly, and may extend into the nearer parts of the mantle canals (Fig. 10). In at least one hermaphrodite (*Platidia*) the gonads ripen at different times, thereby avoiding self-fertilisation,[3] but in another (*Pumilus*) they appear to ripen synchronously.[67] The gonads of brachiopods are sometimes tied to the inner surfaces of the valves by small pillars of connective tissue (Fig. 10A): traces of these are occasionally preserved in fossil brachiopods as areas of small pimples or pits.

Some brachiopods seem to have no special breeding season; others spawn during a more or less limited part of the year.[67] Lingulids breed all the year round in the tropics, but seasonally in temperate waters: this suggests a temperature control.[23, 64] Like other sessile invertebrates, brachiopods may have some chemical mechanism for coordinating the release of sperm and ova. When spawning occurs, the sex cells break loose from the gonads and are drawn by ciliary currents through the coelom into the nephridia. These act as gonoducts, and eject the sperm and ova into the mantle cavity. In most living brachiopods they are then carried to the exterior in the exhalant current, and fertilisation takes place in the sea-water outside. In the lingulide *Glottidia* it has been estimated that about 10,000 ova are shed from a female in the first spawning, and a total of about 130,000 ova in the whole productive lifetime.[64] Comparably large numbers of ova are probably produced by other species employing external fertilisation.

Brood pouches

In a few brachiopods, however, the ova are retained within the mantle cavity of the female, and are fertilised there by sperm carried in with the inhalant current. The number of ova produced may be very small. The larvae then pass their earliest stages of growth within the mantle cavity.[67] In two living genera they are held during this period in special brood pouches. *Argyrotheca*, a small terebratulide, has a pair of pouches formed by modification of the nephridia.[8] *Lacazella* (a thecideacean) has a single median pouch on the ventral side of the body, just behind the mouth; a pair of modified filaments project into the pouch and provide attachment for the larvae (Fig. 89). These filaments leave a pair of small notches in the shelly ridge

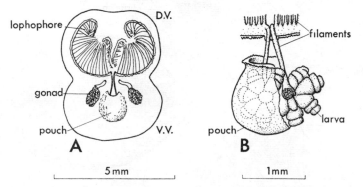

Fig. 89. Brood-pouch in a living strophonenide (*Lacazella*; Thecideacea): A, shell opened, showing pouch in ventral mantle (filaments contracted); B, enlargement of pouch, showing three-segment larvae attached to special lophophoral filaments.[57]

bordering the brachial axis (Fig. 78A); similar notches have been recognised in some—presumably female—specimens of a related genus (*Bifolium*) of early Cretaceous age.[38] This is perhaps the clearest evidence of sexual differentiation known among fossil brachiopods.

Pouches of a different kind are found in some other extinct genera, and may also have been brood pouches. They were formed by a modification of the valves themselves. Postero-laterally, the valve edges were gradually invaginated during ontogeny to form a

pair of shelly pouches open to the exterior (Fig. 90A, B). The inner
surfaces of these pouches might have provided a sheltered place of
attachment for the larvae during their earlier stages of growth (Fig.
90C). These pouches were evidently evolved at least twice: they
developed in the Devonian period in one species of the atrypide
Uncites[56] and again in the late Carboniferous period in the small
rhynchonellide *Cardiarina*.[29] Since they occur on every specimen,
this is not a case of sexual dimorphism, and these brachiopods may
have been hermaphrodite. The pouches are closely analogous in
structure to the single shelly brood pouch of a living bivalve mollusc
(*Thecalia*) which is probably hermaphrodite.[80]

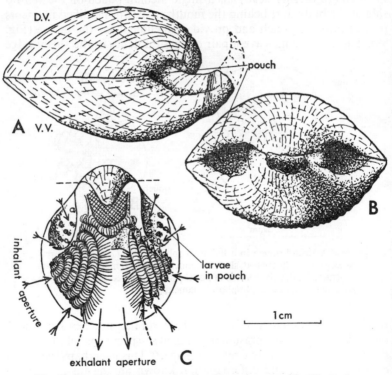

Fig. 90. Possible brood pouches in a Devonian atrypide (*Uncites*).

A, B, lateral and posterior views of shell, showing pouches formed
by invagination of valve edges. C, reconstruction of dorsal valve,
showing possible use of pouches as temporary shelter and anchorage
for larvae. Note that the lophophore would draw fresh inhalant water
through the pouch.[80]

Larval development

Details of embryology and larval development are known only for a very few living species, and then only imperfectly. There seem to be important differences between the inarticulates and the articulates; but since these processes cannot be studied in the extinct brachiopods it is quite uncertain how fundamental the differences really are. They were formerly regarded as extremely significant, being used even to divide the phylum in two, but it is possible that some of the differences are due to larval adaptation.

Almost all accounts of the early embryology date from the last century, and urgently need revision. Descriptions of the early cleavage and the later formation of the mesoderm, for example, leave many uncertainties. Inarticulates are said to be schizocoelic and articulates enterocoelic, but there are conflicting and anomalous reports. At a rather later stage, however, the appearance of the larvae is well known. The inarticulate larva consists of two segments, the future body and the future mantle, whereas the articulate larva consists of three, the future body, mantle and pedicle (Fig. 91A). Parts of the segments are covered with cilia. In articulates the mantle segment at first grows down like a skirt around the pedicle, and later undergoes a sudden *reversal* so that it envelops the body segment.[65, 67] In inarticulates the mantle grows upwards from the start, and there is no reversal.[113] This difference too has been regarded as fundamental, but seems to have no necessary connection with the type of pedicle, and may be of quite minor importance.

The larvae of some species are common elements of the zooplankton at certain seasons. Some are known to be positively phototactic at first,[64] their reaction to light being reversed near the time of spatfall. The articulate larvae have a very short free-swimming period —probably a few hours or at the most a few days—so that the power of dispersal between one generation and the next must be limited. But like the larvae of many other sessile invertebrates, they may be able to delay settlement if the substrate first sampled is unsuitable, or if the water is too deep for the bottom to be reached. In genera with brood pouches most of the 'free' period must be spent in the pouch; this must reduce the power of dispersal still further. But it may increase the chance of survival, for the larvae probably suffer their greatest mortality while they are still part of the plankton. Like the articulates, *Crania* settles at an early stage of growth; lacking a pedicle segment it becomes attached directly by cementation of the ventral valve.[69]

Other living inarticulates have developed a much longer free-swimming period—in *Glottidia* a minimum of three weeks—and they are much further advanced when they settle to the bottom. By this time the valves have begun to grow from the larval protegula; the lophophore is quite well developed and is used for both feeding and swimming; and a pedicle has grown from the posterior mantle edge, ready to be protruded when the larva settles (Fig. 91B).[64]

Rate of growth and population structure

Very little is known about the rate of growth of living brachiopods, their length of life, or their rate of mortality. Many living species are known to reach sexual maturity while the shell is still much smaller than the so-called 'adult' size, so that they must continue to breed for a large part of the total life cycle. In the small lingulide *Glottidia* and the small terebratulide *Pumilus* the average life span is estimated at twenty months and three years respectively;[64, 67] larger brachiopods almost certainly live for several years.[119] On some shells the growth-lines are accentuated at fairly regular intervals: if these represent seasonal periods of retarded growth, they indicate a life span of several years with a gradually slowing rate of growth.[99] Living populations often have a curiously small proportion of young individuals, which may be due to a high rate of mortality in the earlier stages of growth, as in many other animals. If this is largely due to predators, it would explain why most assemblages of fossil species also have a low proportion of small shells, even where there is no evidence that they were drifted away by currents after death.

Even a cursory study of living populations shows that many species have a high degree of true intra-specific variation, in addition to the apparent variation of individuals at different stages of growth. This fact has only recently begun to be appreciated generally by palaeontologists, and there is little doubt that many so-called fossil 'species' are no more than intra-specific variants or even juvenile growth stages. Intra-specific variation is often as marked in so-called 'internal' characters as in the external form and 'ornament' of the shell; and recent studies have shown that this is also true of fossil species,[40] thereby undermining the rather naive confidence that some systematists have had in the stability of 'internal' characters.

Ecological distribution

Living brachiopods are notably patchy in their distribution. Where they occur at all, they may be extremely abundant, closely crowded together and even attached to one another (Fig. 1). Yet in nearby

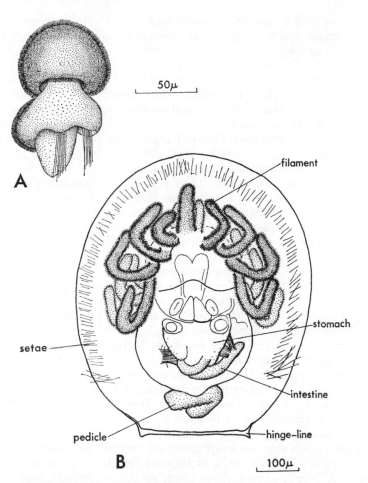

Fig. 91. Free-swimming larvae at growth stages shortly before settlement. A, three-segment larva of a terebratulide (*Waltonia,* Terebratellacea).[65] B, mature larva of the inarticulate *Lingula*: organs seen by transparency through shell; note ciliated filaments of lophophore, blind-ended intestine, and pedicle not yet protruded.[113]

areas of apparently the same character they may be rare or absent altogether.[76] Sometimes this may be controlled by subtle environmental differences, but more generally it is probably due to the limited power of dispersal of the larvae.

This apparent handicap may, however, have been an important

factor in their evolution. It would lead very readily to the spatial, reproductive and hence genetic isolation of parts of the total population of a species; and these are the conditions in which speciation is most likely to occur. On the other hand, despite a patchy distribution, many living species have a very wide total range (e.g. *Terebratulina retusa*, throughout the northern hemisphere). Even with a short free-swimming stage a species could probably spread very rapidly— at least on the geological time-scale—provided that ecological conditions were favourable. Other living species have an extremely narrow known geographical range; they may be relatively newly evolved, or final relics of an old species, or merely species with narrow ecological tolerances. There is so little critical modern work on the ecology of living brachiopods that no more can usefully be said. But certainly there are extinct species with similar patterns of geographical distribution, those that were extremely widespread and those that are known only from a single locality; it is unlikely that the latter are due wholly to the imperfection of the fossil record.

Almost all living brachiopods are stenohaline, being strictly confined to water of full marine salinity. Only the lingulides are known to be able to tolerate brackish water; and in many localities even this may be only a passive toleration, the shell being tightly closed and withdrawn into the burrow until the fully salt water returns with the following tide. Experiments do show, however, that *Glottidia* at least is genuinely euryhaline.[64] Lingulides commonly occur in strata with few if any other brachiopods; they may have lived in similar conditions of variable or reduced salinity.[41] Where fossil articulate brachiopods are conspicuously absent from an apparently marine fauna, it is generally taken to imply that the fauna lived under such conditions,

Factors such as water temperature are probably important for the distribution of brachiopods, as they are for other marine animals, but very little is known about them. At the present day articulate brachiopods and the calcareous inarticulates (craniaceans) are most abundant in temperate waters, but this may not have been so in the past (Fig. 92). Only a few living species, notably the chitinophosphatic inarticulates, are characteristic of tropical or subtropical regions, and the distribution of *Glottidia* seems to be limited by cold winter temperatures.[64]

The range of depth at which living species are found is often surprisingly wide. Some live exclusively in shallow water, a few (e.g. *Pelagodiscus*, exclusively in abyssal depths (none are known from the deep oceanic trenches); but many range from very shallow water into depths of several hundred metres. A very few (e.g. *Discinisca, Waltonia, Terebratalia*)[6, 64, 76] extend just above low tide level, but cannot tolerate more than short periods out of the water. Different assemblages of contemporary Silurian brachiopods have been used to plot a series of probable depth zones parallel to the Silurian shoreline in Britain.[116] In general, fossil brachiopods are most abundant in sediments that are thought to have accumulated in fairly shallow waters.

Many living brachiopods can tolerate quite muddy water, owing to their highly developed mechanisms for rejecting excess mud and

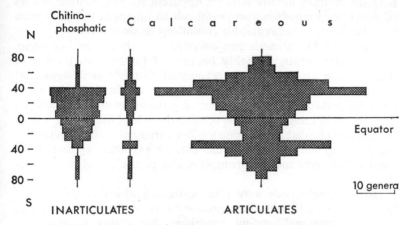

Fig. 92. Latitude distribution of genera of living brachiopods, showing predominately tropical-subtropical occurrence of chitino-phosphatic inarticulates, and predominately temperate occurrence of calcareous shells (including the inarticulate Craniacea).[34]

silt from the lophophore and mantle cavity.[76] They are not able to cope so well with an environment in which mud or silt is being actively deposited. They have to keep the valves free from sediment by frequent snapping of the shell; but there are limits to the efficacy of this mechanism, especially for small and young individuals. Nevertheless, some living species do colonise muddy bottoms; and judging from the abundance of fossil brachiopods in fine grained sediments, this habitat was exploited even more in the past. For living species, the colonisation of muddy bottoms is probably limited chiefly by the available modes of anchorage. Few can root themselves directly into such a sediment, and there may be no other materials available for attachment. Among fossil brachiopods the wider range of available anchoring mechanisms (see Chapter 4), and the much greater development of the free-lying modes of life, seem to have made substrates of soft sediment far more amenable to colonisation. In such environments are found the largest brachiopods that have ever existed (*Gigantoproductus*). Substrates of clean unconsolidated sand are a notably 'difficult' marine environment, but seem to have been colonised successfully by some extinct brachiopods, most of them perhaps free-lying. The smallest known living brachiopod (*Gwynia*) is a member of the highly specialised interstitial fauna of marine sand.[97] The burrowing lingulids are successful in shallow bottoms of more muddy sand.[120] Apparent stunting in assemblages of fossil brachiopods is always difficult to interpret, but may often have been due to unfavourable conditions of some kind.

Few living brachiopods can colonise strongly current- or wave-swept environments, probably because of the limited strength of their pedicle attachment: the cemented *Crania*[69] is a significant exception. Brachiopods with pedicles are confined to the more sheltered parts of such environments, e.g. the under-sides of boulders. The greater development of the cemented mode of life among extinct brachiopods probably laid open a wider variety of shallow-water and current-swept environments than can be exploited by the living species, but many such environments give poor chances of fossilisation.

Extinct brachiopods were often extremely abundant and varied in the biohermal ('reef-like') environments of the Palaeozoic era, rarely as true reef-forming organisms, but rather as abundant elements of the interstitial reef fauna. In these environments there were often species of exceptionally large size, and during the Permian period bioherms supported some of the largest and most bizarre brachiopods in the history of the phylum.[87] Very few brachiopods live on the coral reefs of present seas, though several are associated with deep-water ahermatypic corals.[10]

A few brachiopods are found in the curious faunas of the 'black shale facies' of the Palaeozoic era. This sediment is thought to have accumulated under poorly oxygenated or even anaerobic conditions. Most of its other fossils appear to have been planktonic or nektonic in habit. The brachiopods, which are generally rare, are very thin-shelled, but it seems improbable that they could float, still less swim. But they may have been pseudo-planktonic, being attached to floating seaweed or other organic materials which would not be preserved in the fossil state. Otherwise it becomes necessary to assume either that they could tolerate the exceptionally unfavourable bottom conditions, or else that they had developed flotation mechanisms such as gas vesicles. In the black graptolite shales of the early Palaeozoic era very small inarticulates are often abundant: these may have been nektonic, but it is possible that they were merely the free-swimming larvae of larger benthonic species living elsewhere.[19]

Biotic relations

Brachiopods are commonly found in association with other sessile suspension feeders such as bivalve molluscs, sponges, ascidians, tube-worms, bryozoans, barnacles and ahermatypic corals (Fig. 1). Fossil brachiopods are often most abundant among similar associated organisms. It is notable that the biohermal brachiopods already referred to are rarely found on true coral reefs: in the Mesozoic era, brachiopods are conspicuously absent from the scleractinian coral reefs, but may be abundant on reefs of calcareous sponges.[39] Brachiopods are rare on modern reefs: the best example (*Frenulina*) has bright colour patterning on the shell, not unlike that of many small reef fish. Fossil brachiopods with traces of colour patterning, either radial or concentric, are known as far back as the Devonian period, chiefly among terebratulides.

Living brachiopods, especially those in shallow water, are often densely encrusted with other organisms, including other brachiopods. This probably reflects the intense competition for settling space in environments where oxygen and planktonic food are abundant.[76] Similar encrusting organisms, or at least those with skeletons capable of fossilisation, are often found on the shells of fossil brachiopods as far back as the Ordovician period.[1] Even a complete covering of the shell is probably no great disadvantage to the brachiopod, so long as it retains the power to open and close the shell. Indeed the incrustation may be a positive advantage, for it certainly helps to camouflage the shell: in present-day environments small and young shells without such incrustation are far more conspicuous than the older encrusted shells. They are in any case the more liable to fall to predators, just because they are small and thin-shelled. Encrusting organisms that were probably commensal have been noted on fossil brachiopods (Fig. 93).

At the present day predators on brachiopods include starfish, crustacea, gastropods and fish, which browse over densely populated surfaces of the substrate, devouring the smaller brachiopods. Ciliates and polychaete worms have been noted as important predators of the vulnerable newly settled larvae. Shell-boring gastropods occasionally attack the larger adult shells, and similar circular holes

F

have been found in the shells of fossil brachiopods as far back as the Ordovician period.[17] Other signs of predators are also preserved. Some fossil shells, also as far back as the Ordovician, have repaired injuries to the valve edges; these may have been due to carnivorous brachyodont fish or nautiloid cephalopods.[88] Starfish may be the most important predators of all, as they are of living bivalve molluscs,

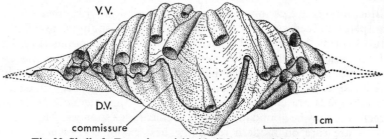

Fig. 93. Shell of a Devonian spiriferide (*Mucrospirifer*) with attached tubes of the commensal organism *Cornulites*.[90] Note that the apertures of most of the tubes are clustered near the lateral parts of the commissure, the inferred positions of the brachiopod's inhalant water currents.

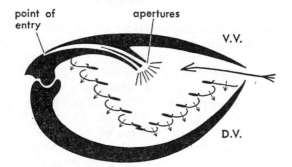

Fig. 94. Reconstruction of the organism *Diorygma*, parasitic on a Devonian *Atrypa*, shown in longitudinal section through one of the lateral spiralia of the brachiopod. The parasite apparently stimulated the brachiopod mantle into secreting a shelly tube, the aperture of which retained throughout ontogeny a position near the centre of the spiralium.[12] The parasite is shown diagrammatically as a suspension-feeder with a crown of filaments, intercepting the normal inhalant water of the brachiopod.[86]

but little is known of their effect on brachiopod populations. Species employing extra-oral digestion may pull the valves slightly apart to insert the stomach into the mantle cavity of the brachiopod: the extremely accurate fit between the valve edges of brachiopods would give some protection against this mode of predation. It may be significant that starfish first appear in the fossil record in the early Ordovician,[20] at about the time that the brachiopods first became very abundant and diverse.

Few parasites have been recorded in living brachiopods, but a trematode and a sporozoan have been found in inarticulates;[62] doubtless more parasites would be found if a search were made for them. The only parasites that can be detected in extinct brachiopods are of course those that affected the preservable shell; but a few are known. A species of *Atrypa*, for example, was infested with an organism (*Diorygma*) which bored through the shell and maintained an aperture within the mantle cavity, near the base of one or both spiralia;[12] in this position the parasite, if it was a suspension-feeder, would have been well placed to intercept the inhalant current of food-bearing water (Fig. 94). Other boring organisms have been reported in the shells of fossil brachiopods,[55] but it is rarely possible to prove that they penetrated the shell while the brachiopod was alive.

8

HISTORY OF THE PHYLUM

Pre-Cambrian origins

In this chapter the evidence of functional evolution given in previous chapters for the various functional systems will be integrated into a tentative reconstruction, in narrative form, of the adaptive evolution of the phylum. The following, final, chapter will discuss some specific problems arising out of this synthesis.

The fossil record unfortunately offers no evidence whatever on the origin of the Brachiopoda or their affinities to other phyla (the evidence of embryology and comparative anatomy is discussed briefly in the next chapter, p. 175). Brachiopods are the first lophophorates, and some of the first metazoans, to appear in the record. Despite intensive search in late Pre-Cambrian strata in many parts of the world, no fossil brachiopods have been found in strata earlier than the Lower Cambrian: all records of alleged Pre-Cambrian brachiopods are now highly doubtful.[43]

Early Cambrian radiation

Early Cambrian brachiopods are often moderately large in size, quite complex in structure, and locally abundant enough to form small shell-banks. They are not notably 'primitive', except by a hindsighted definition of that term. They are also surprisingly diverse. All five orders of inarticulates are represented, and one of the articulate orders. Unfortunately it is difficult to determine the phyletic relations of these groups with any degree of confidence. The morphology of many of them is still poorly known; and the correlation of Lower Cambrian strata is still insufficiently precise to determine the order in which they make their appearance. Clearly the fauna bears witness to a first major phase of evolutionary radiation, and it is not improbable that a large part of this radiation occurred within, or only just before, early Cambrian time.

Phosphatic inarticulates predominate in the fauna, and may represent the original acquisition of mineralised shells. But the more 'difficult' biochemical problem of secreting a calcareous shell was also solved at a very early date (obolellides), perhaps by substitution for phosphate, perhaps independently from small unmineralised ancestors. Some of the inarticulates (acrotretides) have pedicle foramina, though the pedicle may have been of the kind that now characterises the articulates; others (lingulides) have no foramen and may even have lost this primitive pedicle. Most shells have complex muscle scars, suggesting that the inarticulate musculature, with oblique muscles for the adjustment of the valves, was original. But a few (paterinides) have a strophic hinge line and delthyrial covers, which suggest that they had evolved the articulate type of musculature. Other similar shells (kutorginides) are calcareous—probably a separate development of this composition—and only differ from articulates in lacking an internal articulation of teeth and sockets. The first true articulates (billingsellaceans), some of them with only rudimentary articulation, may have been derived from some such ancestors during Lower Cambrian time. They are always relatively inconspicuous elements of the early Cambrian fauna.

Both inarticulates and articulates were distinctly more abundant and diverse by late Cambrian time, but of the early calcareous groups only the true articulates had survived. The only significant addition to the fauna was the appearance of the first pentamerides (porambonitaceans); these differ little from some of the contemporary orthides, and are important chiefly for the variety of the groups that were derived from them subsequently.

Ordovician radiation

Brachiopod-bearing strata of Ordovician age, like early Cambrian strata, are still difficult to correlate accurately on a world-wide scale. But although the details are far from certain, it is already clear that a second and much greater phase of evolutionary radiation occurred in the phylum during the earlier part of this period. Brachiopods became far more diverse and far more abundant than they had been even in later Cambrian times (Fig. 95).

The expansion was most evident in the articulates, which in numbers and variety soon overshadowed the inarticulates. Two new groups of inarticulates developed calcareous shells during the period; one was probably aragonitic and apparently free-lying (trimerellaceans), the other was calcitic and the first to develop a cemented attachment (craniaceans). The appearance of these, together with the far greater expansion of articulates, brought calcareous shells into the majority. But one significant innovation among the phosphatic groups was the development of the burrowing habit among the lingulids. If their ancestors had indeed lost the original larval pedicle segment, they must by this time have acquired a secondary muscular pedicle of the kind on which the burrowing habit depends.

Early in the Ordovician period the expansion of the articulates was most evident in two of the groups that had first appeared in the Cambrian (orthaceans, porambonitaceans). But these were joined by the first of the strophomenides (plectambonitaceans): they had concavo-convex pseudopunctate shells, and were attached by a pedicle (through a supra-apical foramen) only in the early growth stages, becoming free-lying and perhaps quasi-infaunal later in ontogeny. These began to be replaced later in the period by another group (strophomenaceans). At the same time punctate shells appeared for the first time among the orthides (enteletac); and by the end of the period they were beginning to replace the impunctate group (orthaceans) from which they had been derived. Another innovation of great importance was the loss of the hinge line. In some of the early pentamerides this had become vestigial, but the first truly non-strophic shells with a purely internal articulation were rhynchonellides. Likewise some of the early pentamerides may have had rudimentary crura, but these too are much better developed in

Fig. 95. Diversity of the Brachiopoda during geological time, as measured by number of *families* present in specified periods of time: based on Treatise data,[58] but with later additions. 'Present' families are those also present in periods immediately before and after any given period (therefore represented by a rectangle, to denote presence *throughout* the period); 'first' and 'last' families are those first or last known in any given period (therefore represented by a right triangle, to denote their appearance or disappearance *during* the period); 'first and last' families are those only known from a given period (therefore represented by an isosceles triangle, to denote their presence only at some time *within* the period). (Since periods of time used are so unequal, relative proportions of these categories are not equally significant, but are shown in order to indicate derivation of *total* level of graph.) Note low diversity in Cambrian, high level of 'first' families in Ordovician, continuing to peak in Devonian, though balanced by 'last' families; 'plateau' in later Palaeozoic, high level of 'last' families in Permian, low diversity in post-Palaeozoic.[34,86]

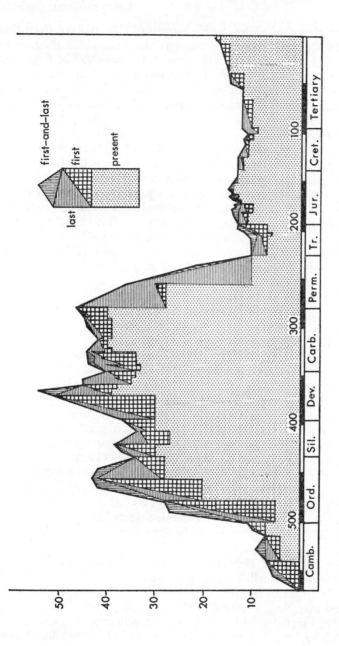

the first rhynchonellides, which were probably evolved from them. The rhynchonellides in their turn were almost certainly the ancestors of the first atrypides somewhat later in the period. The appearance of complex spiral brachidia in this group, involving the power of resorption in the brachidial epithelia, was the last great innovation to appear during the Ordovician period (Fig. 96).

Siluro–Devonian expansion

By the end of the Ordovician period five of the six orders of articulates had emerged as distinct groups, all present in variety and abundance. The following Silurian and Devonian periods were chiefly a time of consolidation. There were few spectacular changes in the fauna, but its overall character altered steadily. Inarticulates were still present at first in some abundance. But with the piecemeal disappearance of several earlier groups, by late Devonian times the inarticulates were virtually reduced to the three families which have subsequently continued to the present day: the pedicle-attached discinids, the cemented craniids and the burrowing lingulids.

Articulates, on the other hand, became steadily more abundant, and probably reached their greatest peak during the Devonian. Strophomenides were particularly abundant, the more 'advanced' continuing to replace the earliest group. Cementation had occurred sporadically among Silurian strophomenides (a few strophomenaceans), but in the Devonian it appeared as a definitive feature of another group (davidsoniaceans). Another important innovation was the first appearance of tubular spines, towards the end of the Silurian period. At first these were probably sensory in function (chonetaceans); but later they began to be developed in greater variety and to be used also for anchorage in soft substrates (productaceans) or for cementation attachment (strophalosiaceans). These groups were, however, relatively unimportant until later in the Palaeozoic.

Meanwhile the punctate orthides continued to flourish, apparently at the expense of the impunctate, and by the end of the Devonian period all the latter had disappeared. Some of the more 'advanced' pentamerides (pentameraceans) became extremely abundant: many of them had large shells, and locally they are preserved in rock-forming quantities. They seem to have acquired crura independently from the rhynchonellides, and many of them also lost their ancestral strophic hinge. In the Devonian period they may have given rise to another group (stenoscismataceans) with rhynchonellide-like shells.

Rhynchonellides and atrypides were also present in rapidly increasing variety. Among one atrypide group (athyraceans) a major

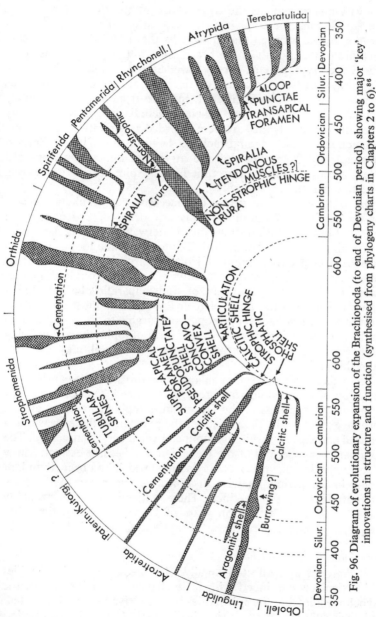

Fig. 96. Diagram of evolutionary expansion of the Brachiopoda (to end of Devonian period), showing major 'key' innovations in structure and function (synthesised from phylogeny charts in Chapters 2 to 6).[86]

G

innovation occurred, in the appearance of the first shells with a transapical foramen, implying the acquisition of the power of resorption by the pedicle (it is noteworthy that resorption was already a feature of the growth of the brachidium in this group). The spiriferides are cryptogenetic, for in lower Silurian strata shells with a long hinge-line (spiriferaceans, cyrtiaceans) appear without obvious ancestors. They are very different in form from the contemporary non-strophic atrypides, and their brachidia are also distinctly different. Tentatively they are here regarded as derivatives from some impunctate orthides, their brachidia being assumed to be derived independently from those of the atrypides. Punctation appeared independently during Silurian time in both spire-bearing orders (i.e. in retziaceans and suessiaceans).

One of these punctate groups (retziaceans) seems a likely ancestor for the earliest terebratulides. This last order of articulates first appeared in late Silurian times but underwent a spectacular radiation during the Devonian period. The terebratulide loop is likely to have been evolved neotenously from a spiriferide brachidium; its emergence may represent the evolution of a zygolophous lophophore, but some of these early loops are of kinds unparalleled among living terebratulides, and the form of their lophophores must remain uncertain. But a few had 'long loops' comparable to those supporting plectolophes in living species: so this form of lophophore, involving the development of a diaphragm for the median coil, may also date from this period.

Towards the end of the Devonian period this steady expansion and radiation of the phylum was brought to an end. One of the major strophomenide groups of the Devonian (strophomenaceans) almost became extinct, while the variety of the punctate orthides was much reduced and the last surviving impunctate members disappeared. The dominant pentamerides of the Devonian also became extinct. Many of the atrypides (atrypaceans, dayiaceans) which had been abundant earlier in the period likewise disappeared, and many of the early terebratulides became extinct. These extinctions were probably spread over some 10 million years of Upper Devonian time, but they were nevertheless relatively abrupt and far reaching.

Late Palaeozoic expansion

Not all groups were affected equally. The three long-ranging families of inarticulates continued with little change. More important, the rhynchonellides and the spiriferides seem to have been little affected, and their subsequent expansion in the early Carboniferous period may represent a tendency to occupy ecological niches left vacant by the earlier spiriferides and the primitive terebratulides. Similarly the 'advanced' strophomenides with cementation and/or tubular spines seem to have expanded subsequently in place of the earlier free-lying or quasi-infaunal groups without spines.

Thus a distinctive new fauna emerged in early Carboniferous time. It is characterised especially by the great abundance of strongly concavo-convex, probably quasi-infaunal, strophomenides, with spines apparently adapted for anchorage in soft substrates (productaceans). Other strophomenides (chonetaceans, davidsoniaceans, strophalosiaceans) were also important. Spiriferides, atrypides and rhynchonellides were abundant, and the terebratulides that had survived the late Devonian began to expand again; some of these groups were particularly characteristic of the reef-like environments of the time. Only a few punctate orthides remained, and the pentamerides were now represented only by one small group (stenoscismataceans).

Later Carboniferous faunas are not so well known, mainly because the most fully explored parts of the world were occupied at this time by extensive coal-swamp sedimentation, and brachiopods are found only in the strata representing occasional marine incursions. But the fauna probably survived elsewhere without major change, because in Lower Permian strata it reappears, different in generic composition but in broader character unaltered. Terebratulides, rhynchonellides, and spiriferides of many different groups were all abundant, but it is among the strophomenides that the most remarkable developments now occurred. The shells that could be cemented to the substrate by their spines (strophalosiaceans) became as abundant as the similar uncemented spiny shells (productaceans). Some of them even in late Carboniferous time had lost the true commissure between the valves, and in Permian time they developed bizarre shells with the dorsal valve reduced to a recessed operculum (richthofeniids) or slotted plate (lyttoniaceans), which seems to have become enveloped in mantle tissue. This may have been connected with the acquisition or perfecting of rhythmic and other aberrant modes of feeding. These aberrant brachiopods, and several other strophomenides, developed exaggerated coral-like forms at this time, and are most characteristic of reef-like environments.[87] Connected with this is their somewhat restricted geographical distribution: they are more or less confined to the ancient seaway 'Tethys' running from east Asia through to the western Mediterranean, with a westward extension to southern North America. This distribution, and the generic diversity gradients of some other strophomenide families, have been used to argue both for and against theories of continental drift.[94] Unfortunately the evidence is still too incomplete to be conclusive in either direction, but it may well be true that some of these rich and curious Permian faunas were tropical or subtropical in distribution.

Permo–Triassic extinction

This characteristic late Palaeozoic fauna, with its addition of distinctive Permian elements, certainly existed well into Upper Permien time without any important change. The phylum as a whole showed no sign of decline. The more bizarre forms might have been interpreted formerly as signs of 'racial old age', but functional analysis suggests that they were highly adapted to definite modes of life.

Yet at or near the end of the Permian period the phylum suffered the greatest crisis of its history, from which it never fully recovered (Fig. 95). The scale of extinction at this time far outmatched that of late Devonian time (Fig. 99). Among the strophomenides the extinctions affected the fairly long-established free-living and quasi-infaunal groups (davidsoniaceans, chonetaceans, productaceans, strophalosiaceans) as drastically as the more recently evolved aberrant forms. In fact the whole strophomenide order, once so prolifically abundant and diverse, disappeared almost completely, leaving only a very few rare survivors, and those so altered that their affinities are still a matter of controversy. The last orthides (enteletaceans), less abundant in Permian time but nevertheless important and distinctive elements of the fauna, disappeared without trace. Likewise the last pentamerides (stenoscismataceans) disappeared, although they too had been quite important in Permian time, and had even developed some striking morphological features. Some important spiriferide groups (spiriferaceans, reticulariaceans, cyrtiaceans) and the last of the primitive terebratulides (stringocephalaceans) also disappeared. For the rhynchonellides the changes may be less real than current taxonomy suggests, for the study of Palaeozoic and Mesozoic forms by separate systematists tends to produce an artificial discontinuity. Likewise there may be more continuity in the terebratulides than the taxonomy would indicate. But even allowing for this, the magnitude of extinction is spectacular.

What is more controversial is its timing. The Permo–Triassic extinctions, which affect not only the brachiopods but also many other phyla of marine invertebrates, have been regarded in the past as an exceptionally sudden or 'catastrophic' event. More recent work on the zonation and correlation of the strata concerned suggests that where sedimentation was most nearly continuous there was in fact a gradual impoverishment of the brachiopod fauna. But in a geological context 'sudden' and 'gradual' are relative terms. What is now clear is that the last known records of many groups of brachiopods are in very late Permian (Dzjulfian) strata (there are even a few controversial reports from strata that are said to be very early Triassic in age). But brachiopods are extremely rare in Lower Triassic strata throughout the world, and only reappear in moderate abundance in the Middle Triassic. Unless undue weight is to be placed on purely negative evidence, it now seems probable that most of the extinctions occurred during a relatively short interval (perhaps only a few million years) of very late Permian and very early Triassic time.

Mesozoic and Cenozoic faunas

The fauna that reappeared in the record in Middle and later Triassic strata is thus very different from the Permian fauna. The three long-ranging inarticulate families are the only element to emerge without any important change, but as before they are a very minor part of the fauna. Two rare groups of what are probably aberrant strophomenides are known (koninckinids, thecospirids); curiously, both of them had acquired spiral brachidia independently, and have been regarded as aberrant spiriferides. A third equally rare group (thecideaceans) is also probably strophomenide in affinities, although it is punctate. It may be significant that all these relict strophomenides are small in size. Apart from these, the Triassic fauna consists of rhynchonellides, atrypides, spiriferides and terebratulides, generally in that order of abundance. The atrypides and spiriferides are much reduced in variety, and the terebratulides mostly belong to new groups. Even allowing for the fact that Triassic brachiopods have received little study in modern times, there is no doubt that the phylum was very much reduced from its former importance.

By the middle of the Jurassic period two of the three surviving strophomenide groups had disappeared, and the third had become extremely rare. More important, all the atrypides and spiriferides had become extinct. Terebratulides were now more abundant than rhynchonellides, and these two orders constituted almost the entire fauna.

From this time until the present day, the character of the faunas has changed little. The end of the Mesozoic era, which marked a time of crisis in several other phyla, had scarcely any effect on the brachiopods except at the generic level. This is in striking contrast to the crisis at the end of the Palaeozoic.

The three persistent groups of inarticulates have survived with little change. The last surviving strophomenides (thecideaceans) had a minor revival in late Cretaceous time, but are now once more extremely rare; and the rhynchonellides seem to have declined steadily in importance (this is somewhat masked by unequal taxonomic treatment). The terebratulides with short loops (terebratulaceans) have remained fairly steady. Only those with long loops (terebratellaceans) have shown any real signs of evolutionary diversification in Cenozoic time, with the appearance of very varied brachidial loops, and the development of forms with simple or lobate lophophores (megathirids). But this structural and functional radiation is on a very minor scale compared with the earlier history of the phylum.

It is interesting to note that there are very few living genera that have not also been recognised in the fossil state: this strongly suggests that, for brachiopods at least, the fossil record may not be as imperfect as it has become fashionable to assume.

9

ASPECTS OF EVOLUTION

Affinities to other phyla

The comparative anatomy of living brachiopods and members of
other phyla has long suggested an affinity with the ectoproct bryozoans, and, more recently, with the phoronids. Similarities include
the lophophore, the trochophore-like larvae, the development of the
mouth from the blastopore, the location of the nerve ganglia in the
mesosome and the lack of a clearly differentiated protosome. But
brachiopods differ in their bivalved mineralised shell, their chitinous
setae and their open circulatory system, and perhaps also in their
metanephridia and their radial cleavage. Nevertheless the similarities
have seemed strong enough to warrant grouping these phyla together
as 'Tentaculata' or 'Lophophorata'.[51] It is unlikely that all the similarities are due to convergence.

If the validity of some such super-phylum is accepted, it is still
difficult to determine its relation to other metazoans. Affinities have
been suggested both with the arthropod-mollusc-annelid group
(Protostomia) and with the echinoderm-chordate group (Deuterostomia). But the brachiopod larva has only a rather general resemblance to the trochophores of Protostomia; there is no definite trace
of cephalisation or metamerism; and the evidence of segmentation
and of the formation of the coelom is equivocal, and could be taken
to suggest affinity with either group. For this reason, it has been
suggested that the lophophorates may have evolved independently
from protozoan ancestors. Alternatively, all three groups may reflect
a very early phase of evolutionary radiation from a single protozoan
stock. We cannot tell how far back in Pre-Cambrian time the first
definitive brachiopods evolved from an original lophophorate stock,
for the fossil record gives no evidence on such speculations. By the

time fossil brachiopods make their appearance they are already quite distinct from any other phylum.

Origin of the Cambrian fauna

There is a real problem here, which needs to be explained and not merely explained away. Late Pre-Cambrian sediments are in many places as unaltered, and as varied in lithology, as later sediments; yet brachiopods, and other metazoans with fossilisable hard-parts, are totally absent. In the Lower Cambrian strata, on the other hand, brachiopods appear in some diversity, and even locally in some abundance. On the total time-scale of the phylum their appearance is curiously abrupt.

Many theories have been proposed to account for this. One explanation is that several metazoan phyla early in the Cambrian period acquired the ability to secrete mineralised skeletons, which greatly increased their subsequent chances of fossilisation.[43] But for brachiopods this poses a further problem, for their entire functional organisation depends on their possession of a rigid shell. It is difficult to conceive that a soft-bodied brachiopod would be recognisable as a brachiopod at all—unless it was small in size. But at a small size a shell composed exclusively of organic material might have been sufficiently rigid. No substantial increase in body size would then be possible until the mantle acquired the ability to secrete mineral material to strengthen the chitin. But until that time brachiopods could have existed in a form that was functionally and structurally similar to the early post-larval stages of living species with small shells composed exclusively of organic material (perhaps homologous to the periostracum),[107] and consequently with very low chances of fossilisation. In such a form brachiopods might have had either a long or a short Pre-Cambrian history, but no trace of it would be preserved.

Some further explanation may be required, however, to account for the acquisition of mineral skeletons by several different phyla, all apparently within the span (perhaps 50 million years) of Lower Cambrian time. The relatively sudden evolutionary episode which gave rise to this Cambrian metazoan fauna may have been 'triggered off' by the climatic amelioration at the end of an infra-Cambrian glaciation.[81] For in many parts of the world there is evidence of a long period of glaciation late in Pre-Cambrian time, and this appears to have been more severe and more widespread than any later Ice Age.

Adaptive radiation

The general course of evolution of the Brachiopoda, at least in its earlier phases, resembles that of several other metazoan phyla (Fig. 96). The earliest phase of radiation, in the Cambrian period, led to structural diversity almost as great as at any subsequent time (this is recognised taxonomically by the number of orders present), although few of the divergent forms were abundant numerically. This pattern is paralleled, for example, in the early history of the molluscs and the echinoderms. It is not yet possible to identify confidently the adaptive elements in this earliest phase of brachiopod radiation; indeed it is possible that in the unusual ecological conditions of the Cambrian seas (e.g. in the apparent absence of significant predators) the radiation may have occurred under conditions of exceptionally low selection pressure.

This is suggested also by the early extinction of several of the divergent groups, the survivors being those responsible for all the subsequent proliferation of the phylum. Thus there emerged only two clear-cut structural types, with complementary adaptive advantages: the presence or absence of true articulation, with its correlated effects on the musculature, giving more effective closure of the shell or else a greater freedom of movement to the valves.

The second phase of radiation, chiefly during the Ordovician period, can also be paralleled in the history of other phyla. The dominant skeletal material switched from the organic-phosphatic to the calcareous. Indeed the expansion of the phylum at this time may have been due in part to the ability of several groups to utilise carbonate in place of the much scarcer phosphate material. In both articulates and inarticulates the burgeoning structural diversification (which is reflected in the taxonomy) has obvious adaptive elements. Indeed, detailed studies of the functional morphology of individual forms is demonstrating increasingly the pervasive adaptive dimension in this radiation.

A crucial role in the radiation was played by a small number of 'key' structural innovations, such as the punctate shell structure, the non-strophic hinge, the concavo-convex shell-form, cementation, the brachidium, complex brachidia involving resorption, and the supra-apical and transapical types of pedicle foramen, and (probably) tendonous muscles. The appearance of these 'strategic' features has rightly been given a prominent place—however implicitly—in the taxonomy of the phylum. It is notable that all of them were evolved for the first time within a mere 80 million years (Lower Ordovician to Lower Silurian inclusive).

Almost certainly the radiation was influenced by ecological interaction with simultaneous developments in other phyla, but it is not yet possible to identify more than a few such possible correlations. The emergence of probably carnivorous organisms early in the Ordovician period (e.g. asteroids and nautiloids) must surely have affected brachiopod faunas in a significant way; but at present we can only hazard tentative guesses that structural features such as valve interlocking, and the development of the quasi-infaunal and true burrowing habits, might have been adaptive reactions to the appearance of such predators. The emergence of abundant jawed fishes in marine environments in later Devonian time may perhaps be reflected in the increasingly dominant role of the quasi-infaunal strophomenides in later Palaeozoic faunas.

Another notable feature of brachiopod radiation is the prevalence of repetitive evolution. It is now clear that few even of the 'strategic' structural features were evolved only on one unique occasion; and at a lower level of structural innovation (e.g. in overall shell-form) the evolution of the same feature again and again is a commonplace. This can be interpreted most coherently as a natural consequence of the basic structural simplicity and conservatism of brachiopod organisation. Given the basic elements of the brachiopod body, and its apparently limited anatomical potentialities, there would be an inherent limitation on the possible variety of modifications that could serve any given function, and hence an inherent likelihood that each available functional solution will have been adopted on several occasions by different groups of brachiopods (Fig. 97).

Homoeomorphy, the extreme form of evolutionary convergence, undoubtedly occurs in the phylum (the term was originally coined for use in brachiopod studies), and in a few cases it can be interpreted in adaptive terms (Fig. 98). But the frequency of its occurrence, as a seriously confusing element in taxonomic studies, has almost certainly been exaggerated, especially by the assumption that 'internal' structural features are less liable to display convergence than the 'external'. However often a given single feature may have been evolved, the brachiopods in which that feature occurs are rarely similar enough in other features to cause serious confusion.

Fig. 97. Diagram to show possible phylogeny of one small group of brachiopods (Thecideacea; Strophomenida); to illustrate repetitive evolution of complex from simple lophophores, governed by functional factors related to absolute size, and also possible neotenous origin of whole group. Each drawing shows a dorsal valve with inferred course of brachial axis (based on grooves in Thecideacea, spiralia in *Thecospira*).[84]

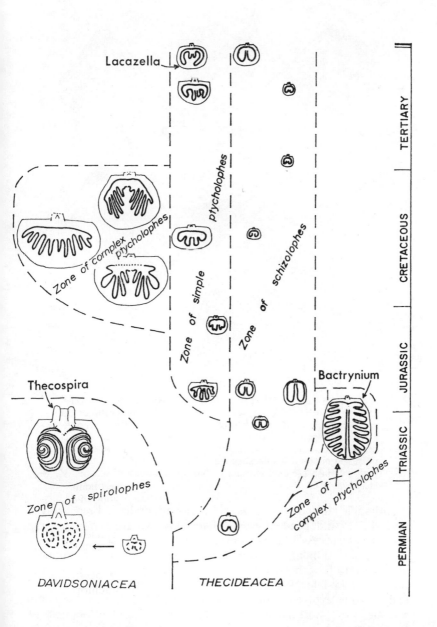

Lacazella

Thecospira

Bactrynium

Zone of complex ptycholophes

Zone of simple ptycholophes

Zone of schizolophes

Zone of spirolophes

Zone of complex ptycholophes

DAVIDSONIACEA

THECIDEACEA

TERTIARY

CRETACEOUS

JURASSIC

TRIASSIC

PERMIAN

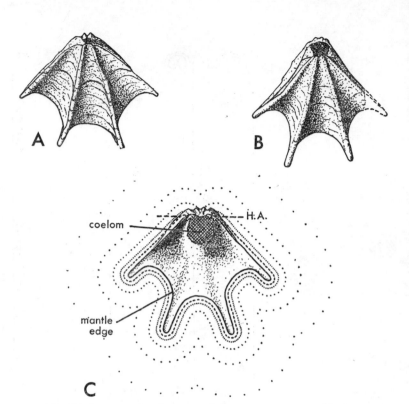

Fig. 98. An example of close evolutionary convergence ('homoe-omorphy') in brachiopods, with a functional interpretation of the resemblance.[82] A, dorsal view of *Tetractinella*, a Middle Triassic atrypide; dorsal view of *Cheirothyris*, an Upper Jurassic tere-bratulide; C, internal view of dorsal valve of either genus, to show possible function of 'spikes' in extending the sensitive mantle edge outwards into the environment (arbitrary 'contours of sensitivity' as in Fig. 60).

Although most of the 'strategic' innovations have been evolved several times independently by different groups of brachiopods, the subsequent 'success' of the groups in which they occurred is curiously uneven. Thus tubular spines, first developed in Silurian strophomen-ides, seem to have been a partial cause of the great proliferation of that order in later Palaeozoic time. But the structurally identical feature, when evolved in Jurassic rhynchonellides, characterised only a couple of short-lived genera. Similarly, the development of brachidia involving resorption seems to have occurred at least four times: but whereas the first time, in Ordovician atrypides, led to all the abundant later atrypides and to all the terebratulides, and the second Silurian occasion led to all the spiriferides, two other developments of the same feature in Triassic strophomenides led only to a few rare short-lived genera (Fig. 88).

The problem of extinction

On the whole, the pattern of extinction shown by the brachiopods in the earlier part of their history is no great problem. Much of it can be interpreted plausibly in terms of the ecological replacement of earlier groups by later groups with demonstrable adaptive advantages. For example, among the concavo-convex strophomenides, the later forms possessing tubular spines (productaceans) were clearly at an advantage for a quasi-infaunal life in soft substrates, relative to the earlier spine-less superfamilies (plectambonitaceans, strophomenaceans) which they seem to have replaced in the same habitat.

The drastic decline of the phylum at the end of the Palaeozoic era is another matter. This is part of a larger problem, since the Permo–Triassic extinctions were equally drastic and—on the geological time-scale—equally sudden in several other phyla of marine animals. This phase of extinction in the Brachiopoda is striking enough when viewed taxonomically, especially in the almost complete elimination of the Strophomenida; but it becomes much more significant when viewed in structural and adaptive terms. For these strophomenides seem to have diverged further than any other group, in both structural and functional terms, from the original articulates. They were probably better adapted than any other articulates, both to an almost infaunal mode of life and also to an epifaunal habit with strong cementation attachment and anchoring spines. Some may have been able to swim, or at least to move rapidly to avoid predators. Many had probably developed rhythmic feeding mechanisms that released them from some of the limitations of ciliary suspension feeding, and some may have brought these mechanisms to a high degree of perfection, even utilising the mantle tissue as a site for food-collection and possibly for symbiotic algae. Many had abandoned the original tooth-and-socket articulation in favour of a more friction-free hinge. Some had also abandoned the close commissure between the valve edges, enabling the valves to develop adaptively independently of each other; and in a few the dorsal valve had even become totally sheathed in mantle tissue.

Though some of these innovations are found only in a few forms, as a whole they bear witness to a degree of plasticity unparalleled at any other time in brachiopod evolution. Brachiopod organisation may indeed have been inherently limited; but the late Palaeozoic strophomenides had gone a long way towards breaking out of those limitations, to the extent that some of them would scarcely be recognisable as brachiopods if less aberrant intermediates were not known. It might have been expected that this degree of innovation would provide a key to a renewed phase of adaptive radiation, enabling the phylum to adopt modes of life hitherto unexploited.[87] In fact, these strophomenides were almost wiped out during what now appears to have been an extremely critical period in marine environments.

For there is geological evidence for some kind of exceptional physical situation at this time. In most areas of Permian rocks sedimentation became non-marine in the course of the period, and areas in which marine conditions persisted into Triassic time are extremely rare or even (some would say) unknown. This is unparalleled in any other part of the geological column. Probably it reflects an exceptional degree of emergence of the continents. This would have progressively restricted the shallow epicontinental seas that had provided environments for most of the abundant brachiopod faunas of the Palaeozoic era; and this might have been particularly disastrous for those of the more aberrant Permian brachiopods that were adapted to reef-like environments.

Nevertheless, the scale of brachiopod extinction may require some more radical explanation. For example, the hypothesis that the seas not only withdrew from the continents to an exceptional degree, but that their surface waters also became brackish at this time, would explain the relatively 'catastrophic' extinction of many groups of stenohaline marine organisms, including the brachiopods.[42] Relics of the stenohaline faunas would have survived only in inland seas or other isolated areas of normal salinity. On the mechanism proposed it might have taken most of Lower Triassic time for the salt balance of the oceans to be restored to normality. (It may be significant that lingulids, which are euryhaline at the present day, are usually the only brachiopods to be recorded from Lower Triassic strata.) Only in Middle Triassic time would the relict stenohaline faunas have been able to spread once more into the ordinary epicontinental marine areas. This hypothesis is admittedly speculative, but it does fit the observed facts extremely well as far as brachiopods are concerned.

Brachiopods and bivalves

Other groups of marine animals were affected equally drastically by the Permo–Triassic crisis. But some of them, for example the ammonoids, soon radiated again and recovered their former abundance and diversity. The brachiopods did not: their *lack* of recovery is the essential problem (Fig. 95). They never afterwards played more than a subordinate role in the ecology of marine environments, and in many of those environments they seem to have been virtually displaced by the bivalve molluscs.

If this had been due to direct competition between the two groups, a much more gradual displacement would have been expected. Instead, as already emphasised, the brachiopods showed few signs of decline even in Permian time; and, it may be added, the bivalves show no obvious signs of their later expansion. Much of their great radiation, from Triassic time onwards, seems to have been due to the proliferation of truly *infaunal* suspension-feeding groups. This mode of life was opened up by the acquisition of fused mantle edges and the consequent possibility of siphon development.[93] This 'strategic' innovation may have occurred once or twice previously, in Palaeozoic time, but if so it led to no great evolutionary expansion, perhaps because quasi-infaunal brachiopods were already too firmly entrenched in soft-substrate habitats. But once those brachiopods had been virtually eliminated at the end of the Palaeozoic era, a recurrence of the same innovation among the bivalves could have been exploited immediately and easily, and could have led to the rapid development and diversification of siphonate groups.

Bivalves undoubtedly have certain inherent functional advantages over the brachiopods. All of them have a muscular foot (unless lost secondarily), which provides a means of mobility for which even the muscular pedicle of some inarticulates is only a poor equivalent. Those with eulamellibranch gills have in addition the fusion of gill-filaments, which seems to permit a much higher rate of filtration than the invariably unfused lophophoral filaments of brachiopods. Finally, those bivalves that have evolved fused mantle edges are able to exploit deep-infaunal modes of life which have never really been available to any brachiopods: the possibly quasi-infaunal habit of many strophomenides, and even the shallow-burrowing habit of lingulids, are again poor equivalents.

Nevertheless, in spite of these inherent handicaps, brachiopods seem to have been able to hold their own against the bivalves in a wide variety of environments throughout the Palaeozoic era, and indeed they were often the predominant element in the benthonic faunas of many areas. The bivalves were inherently better placed to exploit a wider variety of habitats than the brachiopods; yet they do not seem to have done so until many of those habitats (especially the shallow infaunal) had been vacated by the brachiopods. For the brachiopods, the decisive turning point seems to have come at the end of the Palaeozoic era. Thereafter, although they were often locally abundant in certain environments, the range of those environments seems to have become much narrower. The reason for this may perhaps lie in the extinction of so many of the most 'promising' or potentially successful brachiopods of the Permian period.

Conclusion

It has not been possible in this book to offer more than tentative suggestions in answer to the questions posed at the beginning. Most of all, this is due to the lack of accurate information about living brachiopods, and to the lack of critically evaluated functional interpretations of fossil brachiopods. This situation will only be rectified if neontologists regain interest in this apparently 'minor' phylum, and if palaeontologists outgrow their role of subservience to stratigraphy and recover a truly biological viewpoint.

Meanwhile, however, it may be concluded that the Brachiopoda, like other invertebrate phyla, seem to have exploited the potentialities of their basic anatomy and physiology with increasing effectiveness during the Palaeozoic era, and that their pattern of evolution conforms to that of other phyla and is explicable in terms of similar mechanisms. But their basic anatomy was inherently limited in its potentialities; and with the destruction—by fortuitous circumstances —of the most 'promising' innovations, which might have broken out of those limitations, the phylum was doomed thereafter to a role of greatly diminished importance in the marine faunas of the world.

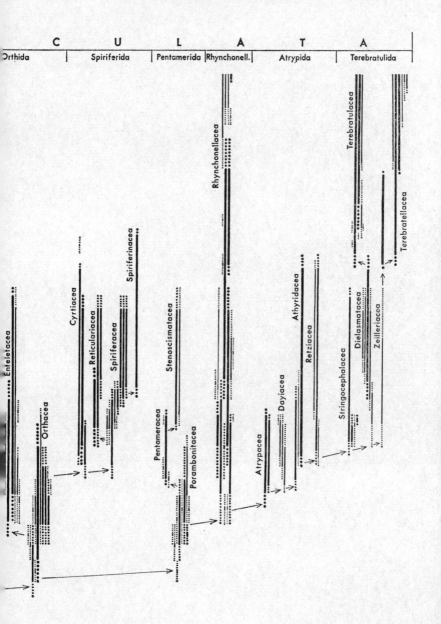

C U L A T A

Orthida | Spiriferida | Pentamerida | Rhynchonell. | Atrypida | Terebratulida

Enteletacea

Orthacea

Cyrtiacea

Reticulariacea

Spiriferacea

Spiriferinacea

Stenoscismatacea

Pentameracea

Parambonitacea

Rhynchonellacea

Atrypacea

Dayiacea

Athyridacea

Retziacea

Stringocephalacea

Dielasmatacea

Zeilleriacea

Terebratulacea

Terebratellacea

REFERENCES

1. AGER, D. V., 1961, The epifauna of a Devonian Spiriferid. *Q. Jl Geol. Soc. Lond.*, **117**, 1–10, pl. 1.
2. ATKINS, D., 1958, A new species and genus of Kraussinidae (Brachiopoda) with a note on feeding. *Proc. zool. Soc. Lond.*, **131**, 559–81.
3. ——, 1959a, The growth stages of the lophophore of the brachiopods *Platidia davidsoni* (Eudes Deslongchamps) and *P. anomoides* (Philippi), with notes on the feeding mechanism. *J. mar. biol. Ass. U.K.*, **38**, 103–32.
4. ——, 1959b, A new species of *Platidia* (Brachiopoda) from the La Chapelle Bank region. *J. mar. biol. Ass. U.K.*, **38**, 133–42.
5. ——, 1959c, The early growth stages and adult structure of the lophophore of *Macandrevia cranium* (Müller) (Brachiopoda, Dallinidae). *J. mar. biol. Ass. U.K.*, **38**, 335–50.
6. ——, 1959d, The growth stages of the lophophore and loop of the brachiopod *Terebratalia transversa* (Sowerby). *J. Morph.*, **105**, 401–26.
7. ——, 1960a, A Note on *Dallina septigera* (Lovén), (Brachiopoda, Dallinidae). *J. mar. biol. Ass. U.K.*, **39**, 91–9, 1 pl.
8. ——, 1960b, The ciliary feeding mechanism of the Megathyridae (Brachiopoda), and the growth stages of the lophophore. *J. mar. biol. Ass. U.K.*, **39**, 459–79.
9. ——, 1961a, The generic position of the brachiopod *Megerlia echinata* (Fischer and Oehlert). *J. mar. biol. Ass. U.K.*, **41**, 89–94.
10. ——, 1961b, The growth stages and adult structure of the lophophore of the brachiopods *Megerlia truncata* (L.) and *M. echinata* (Fischer and Oehlert). *J. mar. biol. Ass. U.K.*, **41**, 95–111.
11. ATKINS, D., and RUDWICK, M. J. S., 1962, The lophophore and ciliary feeding mechanisms of the brachiopod *Crania anomala* (Müller). *J. mar. biol. Ass. U.K.*, **42**, 469–80.
12. BIERNAT, G., 1961, *Diorygma atrypophilia* n. gen., n. sp., a parasitic organism of *Atrypa zonata* Schnur. *Acta palaeont. pol.*, **6**, 17–28, pl. 1–4.
13. BLOCHMANN, F., 1892, *Untersuchungen über den Bau der Brachiopoden* I. Die Anatomie von *Crania anomala* O. F. Müller. Jena. 65 pp., 7 pls.

14. BLOCHMANN, F., 1900, *Untersuchungen über den Bau der Brachiopoden* II & III. Die Anatomie von *Discinisca lamellosa* (Broderip) und *Lingula anatina* Bruguière. Jena. 58 pp., 11 pls.

15. BRUNTON, H., 1964, The pedicle sheath of young productacean brachiopods. *Palaeontology*, **7**, 703–4, pl. 109.

16. ——, 1966a, Silicified productoids from the Viséan of County Fermanagh. *Bull. Br. Mus. (nat. Hist.), Geol.*, **12**, 5, 173–243, pls. 1–19.

17. ——, 1966b, Predation and shell damage in a Viséan brachiopod fauna. *Palaeontology*, **9**, 355–9.

18. BULMAN, O. M. B., 1939, Muscle Systems of some Inarticulate Brachiopods. *Geol. Mag.*, **76**, 434–44.

19. ——, 1964, Lower Palaeozoic Plankton. *Q. Jl geol. Soc. Lond.*, **120**, 455–76.

20. CARTER, R. M., 1968, On the biology and palaeontology of some predators of bivalved Mollusca. *Palaeogeog., Palaeoclimat., Palaeoecol.*, **4**, 29–65, pls. 1–2.

21. CHUANG, S. H., 1956, The ciliary feeding mechanisms of *Lingula unguis* (L.) (Brachiopoda). *Proc. zool. Soc. Lond.*, **127**, 167–89.

22. ——, 1959a, The structure and function of the alimentary canal in *Lingula unguis* (L.) (Brachiopoda). *Proc. zool. Soc. Lond.*, **132**, 283–311.

23. ——, 1959b, The breeding season of the brachiopod *Lingula unguis* (L.). *Biol. Bull.*, **117**, 202–7.

24. ——, 1960, An anatomical, histological and histochemical study of the gut of the brachiopod *Crania anomala*. *Quart. J. Microsc. Sci.*, **101**, 9–18.

25. CHUN, C., 1900, *Aus den Tiefen des Weltmeeres*. Jena.

26. COLEMAN, P. J., 1957, Permian Productacea of Western Australia. *Bull. Bur. Miner. Resour. Geol. Geophys. Aust.*, **40**, 1–189. pls. 1–21.

27. COOPER, G. A., 1930, The brachiopod genus *Pionodema* and its homeomorphs. *J. Paleont.*, **4**, 369–82, pls. 35–7.

28. ——, 1954, Unusual Devonian brachiopods. *J. Paleont.*, **28**, 325–32, pls. 36–7.

29. ——, 1956, New Pennsylvanian Brachiopods. *J. Paleont.*, **30**, 521–30, pl. 61.

30. COPPER, P., 1965, Unusual structures in Devonian Atrypidae from England. *Palaeontology*, **8**, 358–73, pls. 46–7.

31. ——, 1967, Brachidial structures of some Devonian atrypid brachiopods. *J. Paleont.*, **41**, 1176–83, pls. 155–6.

32. COWEN, R., 1966, The distribution of punctae on the brachiopod shell. *Geol. Mag.*, **103**, 269–75.

33. ——, 1968, A new type of delthyrial cover in the Devonian brachiopod *Mucrospirifer*. *Palaeontology*, **11**, 317–27, pls. 63–4.

34. ——, unpublished research.

35. COWEN, R., and RUDWICK, M. J. S., 1966, A spiral brachidium in the Jurassic chonetoid brachiopod *Cadomella*. *Geol. Mag.*, **103**, 403–6.

36. COWEN, R., and RUDWICK, M. J. S., 1967, *Bittnerula* Hall & Clarke, and the evolution of cementation in the Brachiopoda. *Geol. Mag.,* **104**, 155–9.

37. CRICKMAY, C. H., 1950, Some Devonian Spiriferidae from Alberta. *J. Paleont.,* **24**, 219–25, pls. 36–7.

38. ELLIOTT, G. F., 1948, Palingenesis in *Thecidea* (Brachiopoda). *Ann. Mag. nat. Hist.* (12), **1**, 1–30, pls. 1–2.

39. ——, 1950, The Genus *Hamptonina* (Brachiopoda), and the relation of Post-Palaeozoic brachiopods to coral reefs. *Ann. Mag. nat. Hist.* (12), **3**, 429–46, pl. 4.

40. FAGERSTROM, J. A., and BOELLSTORFF, J. D., 1964, Taxonomic criteria in the classification of the Pennysylvanian productoid, *Juresania nebrascensis. Palaeontology,* **7**, 23–8, pl. 2.

41. FERGUSON, L., 1963, The paleoecology of *Lingula squamiformis* Phillips during a Scottish Mississippian marine transgression. *J. Paleont.,* **37**, 669–81.

42. FISCHER, A. G., 1964, Brackish Oceans as the Cause of the Permo–Triassic Marine Faunal Crisis. *Problems in Palaeoclimatology* (Ed. A. E. M. Nairn), 566–74.

43. GLAESSNER, M. F., 1962, Pre-Cambrian Fossils. *Biol. Rev.,* **37**, 467–94, pl. 1.

44. GRANT, R. E., 1963, Unusual attachment of a Permian Linoproductid brachiopod. *J. Paleont.,* **37**, 134–40, pl. 19.

45. ——, 1965, The brachiopod superfamily Stenoscismatacea. *Smithson. misc. Collns.* **148**, 2, 185 pp., 24 pls.

46. ——, 1966, Spine arrangement and life habits of the productoid brachiopod *Waagenoconcha. J. Paleont.,* **40**, 1063–9, pls. 131–2.

47. ——, 1968, Structural adaptation in two Permian brachiopod genera, Salt Range, West Pakistan. *J. Paleont.,* **42**, 1–32, pls. 1–9.

48. HALLAM, A., 1962, Brachiopod life assemblages from the Marlstone rock-bed of Leicestershire. *Palaeontology,* **4**, 653–9.

49. HANCOCK, A., 1859, On the organisation of the Brachiopoda. *Phil. Trans. roy. Soc. Lond.,* **148**, 791–869, pls. 52–66.

50. HARO, A. DE, 1963, Estructura y anatomía comparadas de las gonadas y pedúnculo de los Braquiópodos testicardinos. *Publnes. Inst. Biol. apl., Barcelona,* **35**, 97–117.

51. HYMAN, L. H., 1959, The Lophophorate Coelomates—Phylum Brachiopoda. In *The Invertebrates,* **5** (Smaller Coelomate Groups), 516–609.

52. JAANUSSON, V., 1966, Fossil brachiopods with probable aragonitic shell. *Geol. För. Stockh. Förh.,* **88**, 279–81.

53. JOPE, H. M., 1965, Composition of Brachiopod Shell. In Moore 1965, H156–64.

54. ——, 1967, The protein of brachiopod shell. *Comp. Biochem. Physiol.,* **20**, 593–605.

55. JUX, U., and STRAUCH, F., 1965, Angebohrte Spiriferen-Klappen;

ein Hinweis auf palökologische Zusammenhange. *Senck. leth.*, **46**, 89–125, pls. 7–11.

56. JUX, U., and STRAUCH, F., 1966, Die Mitteldevonische Brachiopod-engattung *Uncites* De France 1825. *Paläontographica*, A, **125**, 176–222, pls. 21–5.

57. LACAZE-DUTHIERS, H. DE, 1861, Histoire Naturelle des Brachiopodes vivants de la Méditerranée. *Annls. Sci. nat.*, (4), **15**, 259–330, pls. 1–5.

58. MOORE, R. C. (Ed.), 1965, *Treatise on Invertebrate Palaeontology*. Part H: Brachiopoda. Univ. Kansas Press and Geol. Soc. Amer.

59. MORTON, J. E., 1960, The functions of the gut in ciliary feeders. *Biol. Rev.*, **35**, 92–140.

60. MUIR-WOOD, H. M., and COOPER, G. A., 1960, Morphology, Classification and Life Habits of the Productoidea (Brachiopoda). *Mem. geol. Soc. Amer.*, **81**, 447 pp., 135 pls.

61. OWEN, G., and WILLIAMS, A., 1969, The caecum of articulate Brachiopoda. *Proc. roy. Soc. Lond.*, B., **172**, 187–201.

62. PAINE, R. T., 1962a, Ecological notes on a Gymnophalline Metacercaria from the Brachiopod *Glottidia pyramidata*. *J. Parasit.*, **48**, 509.

63. ——, 1962b, Filter-feeding pattern and local distribution of the brachiopod *Discinisca strigata*. *Biol. Bull.*, **123**, 597–604.

64. ——, 1963, Ecology of the Brachiopod *Glottidia pyramidata*. *Ecol. Monogr.*, **33**, 187–213.

65. PERCIVAL, E., 1944, A contribution to the life-history of the Brachiopod *Terebratella inconspicua* Sowerby. *Trans. roy. Soc. N.Z.*, **74**, 1–23.

66. RAUP, D. M., 1966, Geometric Analysis of Shell Coiling: General Problems. *J. Paleont.*, **40**, 1178–90.

67. RICKWOOD, A. E., 1968, A contribution to the life history and biology of the brachiopod *Pumilus antiquatus* Atkins. *Trans. roy. Soc. N.Z., Zool.*, **10**, 163–82.

68. RIOULT, M., 1964, Spicules singuliers du Lias de Normandie, analogues aux spicules de certains Brachiopodes actuels. *Bull. Soc. Linn. Normandie*, (10), **4**, 32–6, 1 pl.

69. ROWELL, A. J., 1960, Some early stages in the development of the brachiopod *Crania anomala* (Müller). *Ann. Mag. nat. Hist.*, (13), **3**, 35–52.

70. ROWELL, A. J., and RUNDLE, A. J., 1967, Lophophore of the Eocene Brachiopod *Terebratulina wardenensis* Elliott. *Paleont. Contr. Univ. Kans.*, Paper 15, 8 pp.

71. RUDWICK, M. J. S., 1959, The growth and form of brachiopod shells. *Geol. Mag.*, **96**, 1–24.

72. ——, 1960, The feeding mechanisms of spire-bearing fossil brachiopods. *Geol. Mag.*, **97**, 369–83.

73. ——, 1961a, The feeding mechanism of the Permian brachiopod *Prorichthofenia*. *Palaeontology*, **3**, 450–71, pls. 72–4.

74. RUDWICK, M. J. S., 1961b, 'Quick' and 'catch' adductor muscles in brachiopods. *Nature,* **191**, 1021.
75. ——, 1961c, The anchorage of articulate brachiopods on soft substrata. *Palaeontology,* **4**, 475–6.
76. ——, 1962a, Notes on the ecology of Brachiopods in New Zealand. *Trans. R. Soc. N.Z., Zoology,* **1**, 327–35.
77. ——, 1962b, Filter-feeding mechanisms in some brachiopods from New Zealand. *J. Linn. Soc. (Zool.),* **44**, 592–615.
78. ——, 1964a, The function of zigzag deflexions in the commissures of fossil brachiopods. *Palaeontology,* **7**, 135–71, pls. 21–9.
79. ——, 1964b, The inference of function from structure in fossils. *Brit. J. Phil. Sci.,* **15**, 27–40.
80. ——, 1964c, Brood pouches in the Devonian brachiopod *Uncites. Geol. Mag.,* **101**, 329–33.
81. ——, 1964d, The Infra-Cambrian glaciation and the origin of the Cambrian fauna. In *Problems in Palaeoclimatology* (Ed. A. E. M. Nairn), 150–55.
82. ——, 1965a, Adaptive homoeomorphy in the brachiopods *Tetractinella* Bittner and *Cheirothyris* Rollier. *Paläont. Z.,* **39**, 134–46.
83. ——, 1965b, Sensory spines in the Jurassic brachiopod *Acanthothiris. Palaeontology,* **8**, 604–17, pls. 84–7.
84. ——, 1968a, The feeding mechanisms and affinities of the Triassic brachiopods *Thecospira* Zugmayer and *Bactrynium* Emmrich. *Palaeontology,* **11**, 329–60, pls. 65–8.
85. ——, 1968b, Some analytic methods in the study of ontogeny in fossils with accretionary skeletons. *Paleont. Soc., Mem.,* **2**, 35–59.
86. ——, unpublished research and original figures.
87. RUDWICK, M. J. S., and COWEN, R., 1968, The functional morphology of some aberrant strophomenide brachiopods from the Permian of Sicily. *Boll. Soc. paleont. ital.,* **6**, 113–76, tav. 32–43.
88. SARYCHEVA, T. G., 1949, (Contributions to the study of lesions during the life of shells of Carboniferous productids.) *(Trav. Inst. Paléont. Acad. Sci. U.R.S.S.),* **20**, 280–92, pls. 1–2.
89. SCHMIDT, H., 1937, Zur Morphogenie der Rhynchonelliden. *Senckenbergiana,* **19**, 22–60.
90. SCHUMANN, D., 1967, Die Lebensweise von *Mucrospirifer* Grabau, 1931 (Brachiopoda). *Palaeogeog., Palaeoclimatol., Palaeoecol.,* **3**, 381–92, pls. 1–2.
91. SIMKISS, K., 1964, Phosphates as crystal poisons of calcification. *Biol. Rev.,* **39**, 487–505.
92. SPJELDNAES, N., 1957, The Middle Ordovician of the Oslo Region, Norway. 8. Brachiopods of the Suborder Strophomenida. *Norsk geol. Tidsskr.,* **37**, 1–214, pls. 1–14.
93. STANLEY, S. M., 1968, Post-Paleozoic adaptive radiation of infaunal bivalve molluscs—a consequence of mantle fusion and siphon formation. *J. Paleont.,* **42**, 214–229.

94. STEHLI, F. G., 1964, Permian Zoogeography and its Bearing on Climate. *Problems in Palaeoclimatology* (Ed. A. E. M. Nairn), 537–49.

95. ——, 1965, Paleozoic Terebratulida. In Moore 1965, H730–H762.

96. STEINICH, G., 1965, Die artikulaten Brachiopoden der Rügener Schreibkreide (Unter-Maestricht). *Paläont. Abh.,* A2, 220 pp., 21 pls.

97. SWEDMARK, B., 1959, On the biology of sexual reproduction of the interstitial fauna of marine sands. *Proc. 15th int. Cong. Zool., London,* 327–9.

98. VANDERCAMMEN, A., 1959, Essai d'étude statistique des Cyrtospirifer du Frasnien de la Belgique. *Mem. Inst. r. Sci. nat. Belg., Mém.,* 145, 175 pp., 5 pls.

99. VOGEL, K., 1959, Wachstumsunterbrechungen bei Lamellibranchiaten und Brachiopoden. *Neues Jb. Geol. Paläont., Abh.,* 109, 109–29, Taf. 4, 9 Abb.

100. ——, 1966, Eine funktionmorphologische Studie an der Brachiopodengattung *Pygope* (Malm bis Unterkreide). *Neues Jahrb. Geol. Paläont., Abh.,* 125, 423–40, pls. 38–9.

101. WALLACE, P., and AGER, D. V., 1966, Flume experiments to test the hydrodynamic properties of certain spiriferid brachiopods with reference to their supposed life orientation and mode of feeding. *Proc. geol. Soc. Lond.,* 1635, 160–2.

102. WESTBROEK, P., 1968, Morphological Observations with systematic implications on some Palaeozoic Rhynchonellida from Europe, with special emphasis on the Uncinulidae. *Leid. geol. Meded.,* 41, 1–82, 14 pls.

103. WILLIAMS, A., 1953, North American and European Stropheodontids: their morphology and systematics. *Mem. geol. Soc. Amer.,* 56, 67 pp., 13 pls.

104. ——, 1965, Suborder Oldhaminidina Williams, 1953. In Moore, 1965, H510–H521.

105. ——, 1968a, Evolution of the shell structure of articulate brachiopods. *Spec. Pap. Palaeont.,* 2, 55 pp., 24 pls.

106. ——, 1968b, Shell structure of Billingsellacean brachiopods. *Palaeontology,* 11, 486–90, pl. 91–2.

107. ——, 1968c, Significance of the structure of the brachiopod periostracum. *Nature,* 218, 551–54.

108. ——, 1968d, A history of skeletal secretion among articulate brachiopods. *Lethaia,* 1, 268–87.

109. WILLIAMS, A., and ROWELL, A. J., 1965, [Brachiopod] Morphology. In Moore, 1965, H57–H155.

110. WILLIAMS, A., and WRIGHT, A. D., 1961, The origin of the loop in articulate brachiopods. *Palaeontology,* 4, 149–76.

111. ——, 1965, Orthida. In Moore 1965, H299–H359.

112. WRIGHT, A. D., 1966, The shell punctation of *Dicoelosia biloba* (Linnaeus). *Geol. För. Stockh. Förh.,* 87, 548–56.

113. YATSU, N., 1902, On the development of *Lingula anatina*. *J. Coll. Sci., imp. Univ. Tokyo*, **17**, art. 4, 112 pp., 8 pls.

114. ZIEGLER, A. M., 1966, The Silurian Brachiopod *Eocoelia hemisphaerica* (J. de C. Sowerby) and related species. *Palaeontology*, **9**, 523–43, pls. 83–4.

115. ZIEGLER, A. M., BOUCOT, A. J., and SHELDON, R. P., 1966, Silurian pentameroid brachiopods preserved in position of growth. *J. Paleont.*, **40**, 1032–6, pls. 121–2.

116. ZIEGLER, A. M., COCKS, L. R. M., and BAMBACH, R. K., 1968, The Composition and Structure of Lower Silurian Marine Communities. *Lethaia*, **1**, 1–27.

ADDENDA

117. CHUANG, S. H., 1964, The circulation of coelomic fluid in *Lingula unguis*. Proc. zool. Soc. Lond., **143**, 221–37.

118. COWEN, R., in press, Analogies between the Recent bivalve *Tridacna* and the fossil brachiopods Lyttoniacea and Richthofeniacea. *Palaeogeog., Palaeoclimat., Palaeoecol.*

119. PAINE, R. T., 1969, Growth and size distribution of the brachiopod *Terebratalia transversa* Sowerby. *Pacific Sci.*, **23**, 337–43.

120.——, 1970, The sediment occupied by Recent lingulid brachiopods and some paleoecological implications. *Palaeogeog., Palaeoclimat., Palaeoecol.*, **7**, 21–31.

121. RUDWICK, M. J. S., in press, The functional morphology of the Pennsylvanian oldhaminoid brachiopod *Poikilosakos*. *Smithson. misc. Collns.*

INDEX

Page numbers in italics refer to illustrations

Acanthothiris, 107, *108*
accretion, 30, 41
acrotretaceans, 70, 80, 134
acrotretides (Acrotretida), 21, *22*, 165
adaptive radiation, 177, 182
adductors, 56, *59*, *61*, *63*, *64*, *65*, *67*, *68*
adjustor muscles, *65*, 75
affinities to other phyla, 175
alae, 90
Ambocoelia, 35
amino acids, 34
ammonoids, 183
amphithyrid foramen, 82
ampullae, 39
Anathyris, 55
anchorage spines, *86*, *92*, *93*, 148, 168
anus, *122*
apertures, 118, 124, *126*, 135, 137
apex, *31*, 32
aragonite, 36
aragonitic shells, 165
Argyrotheca, 134, 152, 153
Articulata, 18

articulation, 51, *53*, *54*, *57*, *59*, *60*, *72–3*, 165
ascidians, 77, 117, 161
asteroids (starfish), 162, 178
athyridaceans, 83, 139, 168
Atrypa, *138*, *162*, 163
atrypaceans, 137, 139, 170
atrypides (Atrypida), 21, *23*
Auloprotonia, *35*

Bactrynium, *144*, *179*
barnacles, 161
Bifolium, 153
billingsellaceans, 36, 60, 71, 82, 165
biohermal ('reef-like') environments, 148, 160, 171
biotic relations, 161
bivalve molluscs, 17, 37, 50, 52, 55, 58, 95, 117, 121, 124, 154, 161, 182, 183
black shale facies, 160
blood vessels, 38
body wall, *19*, 20, 37
brachia, 19, 117
brachial axis, 117, *118*

brachial grooves, 130, *131*, 143, *150–1*
brachial pouches, 143
brachial ridges, 135
brachial valve, 20
brachidium, 20, 21, 127, *150–1*
brachiophores, 128
brachyodont fish, 162
brood pouches, *153, 154*
bryozoans, *15*, 132, 161, 175
burrowing brachiopods, *94, 96–7, 124*, 166

CAECAE, *42*
calcareous shells, 34, 165
calcite, 34, 39
camarophorium, 64
Cambrian faunas, 164, 176
Carboniferous faunas, 64, 170
Cardiarina, 154
cardinal areas, 50
cardinal process, *53*, 61, *62–4*, 67, *68*
cardinalia, 52, 76
catch muscles, 57
cementation, 85, *96–7*, 168
Cenozoic faunas, 173
chambers, 117, *119*
Chatwinothyris, 88
Cheirothyris, 110, *180*
Cheirothyropsis, 110
chilidium, 59
chitin, 34
chitinophosphatic shells, 34, 165
Chlidonophora, 78, *79*, 82
chonetaceans, 91, *109*, 168, 171, 172
Chonetes, 91
Chonosteges, 86
cicatrix, 86
cilia, 117, *118*, 121, 130
ciliary pump, 117
circulation, 38
clasping spines, *84*
classification, 21, 26

Clavigera, 55
climatic distribution, 158, *159*
clitambonitaceans, 60, 66, 82
Clitambonites, 81
coelom, *19*, 20, 37, *38*, 123
coelomic fluid, 38
columnar muscles, *64, 66, 72–3*
commensal organisms, 161, *162*
commissural plane, *18*, 32
commissure, *18, 33*
composition of the shell, 34, *48–9*
concavo-convex shell form, 90
Conchidium, 35
corals, 161
costae, 33, 55
costellae, *102*, 103, *104, 114–15*
Crania, 36, 39, *47*, 70, 71, 74, 85, *118*, 121, *122*, 155, 160
craniaceans, 36, 46, 85, 105, 166
craniids, 168
Crenispirifer, 113
Cretirhynchia, 80, 128
crura, 127, *128, 129*, 135, *150–1*, 166
cruralium, 64
Ctenalosia, 54
current-swept environments, 160
current systems, 117, *119*, 132, *136, 138*
Cyclothyris, 111
Cyrtia, 89
cyrtiaceans, 130, 170, 172
Cyrtina, 89
cyrtinids, 66
cytoplasmic threads, 39, 43

Dallina, 129
davidsoniaceans, 46, 66, 85, 168, 171, 172
dayiaceans, 170
defaecation, 122
deflections, 33
delthyrial covers, 59, *60, 61*
delthyrial gap, 58
delthyrium, *56*, 58

deltidial plates, 59, *80*, 81
deltidium, 59, *81*
dental plates, 52
denticles, 52, *53*, 54
depth distribution, 158
deuterolophe, 137
Deuterostomia, 175
Devonian faunas, 46, 47, 55, 168
diductors, *57*, *59*, *61*, *63*, *64*, *65*, *67*, *68*
digestion, 122
digestive diverticula, *65*, 122
Dinobolus, 70
Diplospirella, 139
discinaceans, 80
discinids, 168
Discinisca, 34, 39, *69*, *76*, 78, 135, *137*, 159
diurnal deposition, 40
divaricators, 57
diversity, *167*
dorsal valve, 18, *19*

ECOLOGICAL DISTRIBUTION, 156
Eleutherokomma, *90*
embryology, 155
encrusting organisms, 161
enteletaceans, 44, 166, 172
Eocoelia, 55
epithyrid foramen, 82
Estlandia, 60, *61*, *101*
Eudesia, *133*
excretion, 123
extinction, 181

FAECES, 76, 123
Fardenia, *84*
feeding mechanisms, 120
fertilisation, 152
fibrous layer, 39
filaments, 117, *118*, 130, 183
fold, *33*, 125
follicles, 100, *101*

food-collecting spines, 146, *147*
food groove, *118*, 121
foramen, 20
free-lying brachiopods, 87, 166
Frenulina, 161

GASEOUS EXCHANGE, 39
gastropods, 117
Gemmellaroia, 67, *68*
Geyerella, 35
Gigantoproductus, 160
Gisilina, *83*
Glossinotoechia, 35
Glossinulus, *116*
Glottidia, 39, 152, 156, 158
gonads, *38*, 152
gonambonitids, 41
great brachial canal, *118*, *119*, 126
growth-lines, 31, *33*
growth of the shell, 30, *37*
Gryphus, 65
gut, *19*, *65*, *122*
Gwynia, 132, 160

Hebertella, 59
Hemithyris, *38*, 122
Hesperorthis, 56
hinge, 18, 50, *72–3*
hinge axis, *18*, *35*, 50, *51*
hinge line, 50, *51*, 54, 56, *60*
hinge plate, 52
homoeomorphy, 45, 178, *180*
Homoeorhynchia, *113*
Hustedia, *113*
hydrostatic skeleton, 127, 135

IMPUNCTATE STRUCTURE, 41
Inarticulata, 18
interareas, 50, *51*, *53*, 56, *60*
interlocking buttresses, 55, *56*
intestine, *65*, *122*
intra-specific variation, 156

JUGUM, 139

Kayseria, 139
'key' innovations, *169*, 177
Koninckina, *138*
koninckinaceans, 139
koninckinids, 173
kraussinids, 130, 140
kutorginides (Kutorginida), 21, 36, 70, 165

Lacazella, *64*, 74, 85, *131*, 143, *153*, *179*
lamellae, 128, 137, 140
larvae, 155, *157*
Leptaena, *35*, *41*, 84
leptocoeliids, 139
life span, 156
ligament, 58
Lingula, 14, 31, 34, 39, *69*, 70, 76, 78, *94*, 95, 121, *124*, *157*
lingulaceans, 95
lingulides (Lingulida), 21, 152, 165, 166, 168, 182
logarithmic spiral form, *31*, 32
long loops, *129*, *131*, 143, 170
loops, 24, *65*, *119*, *129*, 140, *150–1*, 170
Lophophorata, 175
lophophore, *19*, 117
lophophore growth, 130, *133*
Lower Cambrian faunas, 24, 36
Lyttonia, 145
lyttoniaceans, 29, *30*, 85, 130, 144, 148, 171

Macandrevia, *38*
Magas, *131*
Magellania, *43*, 99
mantle, *19*, *37*, *41*
mantle canals, *38*, 100, *101*
mantle cavity, *19*, 117
mantle cilia, 121

mantle nerves, 98, *99*
marginal grooves, 100, *101*, *102*, *103*
Marginifera, *92*
Matutella, 83
median coil, 142, 170
median deflection, *33*, 125, *126*, 137
median plane, *18*, 32
median septum, 52, *63*, 64, *67*, 130, *131*
Meekella, 32
megathirids, 55, 105, 174
Megathiris, 143, *144*
Megerlia, 105
Meonia, 120
Mesozoic faunas, 47, 143
Mimella, 45
mouth, *65*, *122*
mucopolysaccharides, 44
Mucrospirifer, 90, *162*
mucus, 95, 121, 124
muddy environments, 159
muscle platforms, *60*, *61*, *62*, *63*, *64*, *66*, *72–3*
muscle scars, 57
muscle systems, *57*, *59*, *63*, *64*, 67', 68, *72–3*

NAUTILOIDS, 162, 178
Nayunella, 111, *113*
Neothyris, 87, *105*, *123*
nephridium, 123
nervous system, *99*
Nisusia, 106
non-strophic hinge, *51*, 54, 55, *72–3*, 166
Notosaria, 15, *37*, *101*, 105, *106*, *118*, *119*, *136*

OBLIQUE MUSCLES, *69*, 70
obolellides (Obolellida), 21, 36, 165
obolides (Obolida), *22*

Obolus, *69*, 70
Obturamentella, *116*
oesophagus, *65*, *122*
Onniella, *103*
ontogeny, 14, *31*
ontogeny of lophophore, *133*
Ordovician faunas, 25, 36, 45, 55, 165
orthaceans, 166
orthides (Orthida), 21, *22*
Overtonia, *63*

Palaeostrophomena, *102*
parasites, *162*, 163
Parenteletes, *113*
paterinides (Paterinida), 21, 70, 165
pedicle, *19*, 20, *65*, *75*, *76*, *157*, 183
pedicle attachment, 77
pedicle foramen, 78
pedicle rootlets, 78, *79*
pedicle sheath, *84*
pedicle structure, 74
Pelagodiscus, 134, 158
pentameraceans, 168
pentamerides (Pentamerida), 21, *23*
periostracum, *37*, 39, *42*, 44
Permian faunas, 47, 171
Permo-Triassic extinctions, 172, 181
phoronids, 175
phosphates, 36, 177
phylogeny, 14, 24, *169*
Pionodema, 45
Platidia, *118*, 152
platidiids, 55
Platystrophia, *113*
plectambonitaceans, 106, 166, 181
plectolophe, *133*, *141*, 142, 170
Plicatifera, *84*
Poikilosakos, 144
polychaete worms, *15*, 77, 161
population structure, 156
porambonitaceans, 165, 166

Pre-Cambrian origins, 164
predators, 161, 178
primary layer, *30*, *37*, 39
prismatic layer, 40
productaceans, 54, 62, 93, 168, 171, 172, 181
Prorichthofenia, *54*
protegula, *31*
protein, 40
Protostomia, 175
Protozyga, *129*
pseudodeltidium, 59, 84
pseudofaeces, 121
pseudopunctate structure, 41, 166
pseudospondylium, 64
ptycholophe, *131*, *133*, 143, *144*, *179*
Pugnax, *113*
Pumilus, *15*, *133*, 134, 152, 156
punctae, 42, *43*, *47*
punctate structure, 41, 42, 166
Pygope, 125, *126*

QUASI-INFAUNAL BRACHIO-PODS, 91, *93*, 96–7, 148, 166, 171, 178, 181, 183
quick muscles, 57, *91*, 100

Rafinesquina, *35*
rejection mechanisms, 121
Renssaeleria, *142*
repetitive evolution, 178, *179*
resorption, 82, 128, *129*, *150–1*, 168, 180
reticulariaceans, 172
retziaceans, 44, 46, 83, 170, 172
Rhactorhynchia, *33*
rhynchonellaceans, 46
rhynchonellides (Rhynchonellida), 21, *23*
Rhynchopora, 112
rhynchoporids, 46
rhythmic feeding mechanisms, 145, *147*, 148, 171

Richthofenia, 35, 147
richthofeniids, 29, *30*, 54, 85, 112, 145, 171
Rimirhynchia, 111

SALINITY TOLERANCE, 158, 182
Scacchinella, 67
schizolophe, *120, 133*, 134, *136, 141*
secondary layer, *30, 37*, 39
sensory spines, *91, 107, 108, 109*, 112, *114–15, 116*, 168
septibranch molluscs, 146
setae, 95, 100, *101, 102, 103, 105, 114–15, 124*
setal grille, 105, *106*
shadow reflex, 98
shell cavity, 19
short loop, 142
Silurian faunas, 46, 168
siphonotretaceans, 80, 84
snapping reaction, 91, 98, *123*, 146
socket plates, 52
sockets, 18, *51, 56, 60*
spicules, 37, 120, 127
spines, *96–7*
spiral brachidia, 137, *138*, 139, 168
spiralia, 137, *150–1*
Spirifer, 138
spiriferaceans, 130, 170, 172
spiriferides (Spiriferida), 21, *23*
spiriferinaceans, 44, 46
spirolophes, *133*, 135, *136, 138*
spondylium, 64
sponges, *15*, 161
spyridium, 64
stegidia, *96–7*
stegidial plates, 60, 82
Stenoscisma, 63
stenoscismataceans, 168, 171, 172
stomach, *65, 122*
Streptis, 33
stringocephalaceans, 140, 172

stringocephalids, 55
strophalosiaceans, 54, 66, 85, 93, 148, 168, 171, 172
Stropheodonta, 53
stropheodontids, 52, 62
strophic hinge, 50, 51, *72–3*
strophomenaceans, 85, 166, 168, 170, 181
strophomenides (Strophomenida), *22*, 24
Strophonelloides, 62
structure of the shell, 39, *48–9*
subapical foramen, *80*, 81, *96–7*
suessiaceans, 44, 46, 82, 170
sulcus, *33*, 125
supra-apical foramen, 83, *96–7*, 166
swimming mechanism, *91*
symphytium, 59, 81
Syringospira, 89

TALEOLAE, *41*, 42
teeth, 18, *51, 56, 60, 65*
tendonous muscles, 64, *65, 66, 72–3*
Terebratalia, 159
terebratellaceans, 130, 134, 140, 143, 174
terebratulaceans, 134, 140, 142, 174
terebratulides (Terebratulida), *23*, 24
Terebratulina, 55, 77, 82, 103, 105, 120, 158
Terebrirostra, 35
Tetractinella, 110, *180*
thecideaceans, 45, 46, 85, 105, 130, 143, 173, 174, *179*
Thecidellina, 134
Thecospira, 139, *179*
thecospirids, 62, 173
trails, 148
transapical foramen, 82, *83, 96–7*, 170
transmedian muscles, 70

Triassic faunas, 46, 55, 173
tridacnid molluscs, 149
trimerellaceans, 36, 71, 80, 165
triplesiaceans, 62
trocholophe, 132, *133*, *136*, *141*
tubular spines, 110, 180

Umbo, 32
Uncinulus, *116*
Uncites, 32, *154*
Upper Devonian extinctions, 170

VALVES, 17
ventral valve, 18, *19*

Waagenoconcha, *93*
Waltonia, *15*, *57*, *75*, *119*, *141*,
    157, 159

ZEILLERIACEANS, 143
zigzag slits, 110, *111*, *113*, *114–15*
zygolophe, *133*, 140, *141*, 170